GRAPH
ALGORITHMS

COMPUTER SOFTWARE ENGINEERING SERIES

ELLIS HOROWITZ, Editor
University of Southern California

CALINGAERT
Assemblers, Compilers, and Program Translation

CARBERRY, KHALIL, LEATHRUM, and LEVY
Foundations of Computer Science

EVEN
Graph Algorithms

FINDLAY and WATT
PASCAL: An Introduction to Methodical Programming

HOROWITZ and SAHNI
Fundamentals of Computer Algorithms

HOROWITZ and SAHNI
Fundamentals of Data Structures

GRAPH ALGORITHMS

SHIMON EVEN
Technion Institute

Computer Science Press

Computer Science Press, Inc.
11 Taft Ct.
Rockville, Maryland 20850

3   4   5   6   85   84   83   82   81

Library of Congress Cataloging in Publication Data

QA
166
E93

Even, Shimon.
  Graph algorithms.

  (Computer software engineering series)
  Includes bibliographies and index.
  1. Graph theory. 2. Algorithms. I. Title.
II. Series.
QA166.E93   511'.5   79-17150
ISBN 0-914894-21-8
UK 0-273-08467-4

PREFACE

Graph theory has long become recognized as one of the more useful mathematical subjects for the computer science student to master. The approach which is natural in computer science is the algorithmic one; our interest is not so much in existence proofs or enumeration techniques, as it is in finding efficient algorithms for solving relevant problems, or alternatively showing evidence that no such algorithm exists. Although algorithmic graph theory was started by Euler, if not earlier, its development in the last ten years has been dramatic and revolutionary. Much of the material of Chapters 3, 5, 6, 8, 9 and 10 is less than ten years old.

This book is meant to be a textbook of an upper level undergraduate, or graduate course. It is the result of my experience in teaching such a course numerous times, since 1967, at Harvard, the Weizmann Institute of Science, Tel Aviv University, University of California at Berkeley and the Technion. There is more than enough material for a one semester course, and I am sure that most teachers will have to omit parts of the book from their course. If the course is for undergraduates, Chapters 1 to 5 provide enough material, and even then the teacher may choose to omit a few sections, such as 2.6, 2.7, 3.3, 3.4. Chapter 7 consists of classical nonalgorithmic studies of planar graphs, which are necessary in order to understand the tests of planarity, described in Chapter 8; it may be assigned as preparatory reading assignment. The mathematical background needed for understanding Chapter 1 to 8 is some knowledge of set theory, combinatorics and algebra, which the computer science student usually masters during his freshman year through a course on discrete mathematics and a course on linear algebra. However, the student will also need to know a little about data structures and programming techniques, or he may not appreciate the algorithmic side or miss the complexity considerations. It is my experience that after two courses in programming the students have the necessary knowledge. However, in order to follow Chapters 9 and 10, additional background is necessary, namely, in theory of computation. Specifically, the student should know about Turing machines and Church's thesis.

v

The book is self-contained. No reliance on previous knowledge is made beyond the general background discussed above. No comments such as "the rest of the proof is left to the reader" or "this is beyond the scope of this book" is ever made. Some unproved results are mentioned, with a reference, but not used later in the book.

At the end of each chapter there are a few problems which the teacher can use for homework assignments. The teacher is advised to use them discriminately, since some of them may be too hard for his students.

I would like to thank some of my past colleagues for joint work and the influence they had on my work, and therefore on this book: I. Cederbaum, M. R. Garey, J. E. Hopcroft, R. M. Karp, A. Lempel, A. Pnueli, A. Shamir and R. E. Tarjan. Also, I would like to thank some of my former Ph.D. students for all I have learned from them: O. Kariv, A. Itai, Y. Perl, M. Rodeh and Y. Shiloach. Finally, I would like to thank E. Horowitz for his continuing encouragement.

S.E.

Technion, Haifa, Israel

CONTENTS

Chapter 1

PATHS IN GRAPHS

1.1 INTRODUCTION TO GRAPH THEORY

A *graph* $G(V, E)$ is a structure which consists of a set of *vertices* $V = \{v_1, v_2, \ldots\}$ and a set of *edges* $E = \{e_1, e_2, \ldots\}$; each edge e is *incident* to the elements of an unordered pair of vertices $\{u, v\}$ which are not necessarily distinct.

Unless otherwise stated, both V and E are assumed to be finite. In this case we say that G is finite.

For example, consider the graph represented in Figure 1.1. Here $V = \{v_1, v_2, v_3, v_4, v_5\}$, $E = \{e_1, e_2, e_3, e_4, e_5\}$. The edge e_2 is incident to v_1 and v_2, which are called its *endpoints*. The edges e_4 and e_5 have the same endpoints and therefore are called *parallel* edges. Both endpoints of the edge e_1 are the same; such an edge is called a *self-loop*.

The *degree* of a vertex v, $d(v)$, is the number of times v is used as an endpoint of the edges. Clearly, a self-loop uses its endpoint twice. Thus, in our example $d(v_4) = 1$, $d(v_2) = 3$ and $d(v_1) = 4$. Also, a vertex v whose degree is zero is called *isolated*; in our example v_3 is isolated since $d(v_3) = 0$.

Lemma 1.1: The number of vertices of odd degree in a finite graph is even.

Proof: Let $|V|$ and $|E|$ be the number of vertices and edges, respectively. Then,

$$\sum_{i=1}^{|V|} d(v_i) = 2 \cdot |E|,$$

since each edge contributes two to the left hand side; one to the degree of each of its two endpoints, if they are different, and two to the degree of its endpoint if it is a self-loop. It follows that the number of odd degrees must be even.

$$\text{Q.E.D.}$$

1

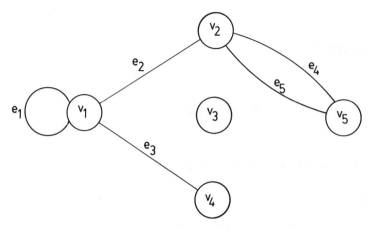

Figure 1.1

The notation $u \xrightarrow{e} v$ means that the edge e has u and v as endpoints. In this case we also say that e *connects* vertices u and v, and that u and v are *adjacent*.

A *path* is a sequence of edges $e_1, e_2, \ldots$ such that:

(1) e_i and e_{i+1} have a common endpoint;
(2) if e_i is not a self-loop and is not the first or last edge then it shares one of its endpoints with e_{i-1} and the other with e_{i+1}.

The exception specified in (2) is necessary to avoid the following situation: Consider the graph represented in Figure 1.2.

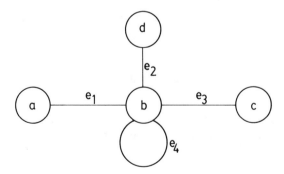

Figure 1.2

We do not like to call the sequence e_1, e_2, e_3 a path, and it is not, since the only vertex, b, which is shared by e_1 and e_2 is also the only vertex shared by e_2 and e_3. But we have no objection to calling e_1, e_4, e_3 a path. Also, the sequence e_1, e_2, e_2, e_3 is a path since e_1 and e_2 share b, e_2 and e_2 share d, e_2 and e_3 share b. It is convenient to describe a path as follows: $v_0 \overset{e_1}{-} v_1 \overset{e_2}{-} v_2 \cdots$ $v_{l-1} \overset{e_l}{-} v_l$. Here the path is e_1, e_2, $\ldots$, e_l and the endpoints shared are transparent; v_0 is called the *start* and v_l is called the *end* vertex. The *length* of the path is l.

A *circuit* is a path whose start and end vertices are the same.

A path is called *simple* if no vertex appears on it more than once. A circuit is called *simple* if no vertex, other than the start-end vertex, appears more than once, and the start-end vertex does not appear elsewhere in the circuit; however, $u \overset{e}{-} v \overset{e}{-} u$ is not considered a simple circuit.

If for every two vertices u and v there exists a path whose start vertex is u and whose end vertex is v then the graph is called *connected*.

A *digraph* (or *directed graph*) is defined similarly to a graph except that the pair of endpoints of an edge is now ordered; the first endpoint is called the *start-vertex* of the edge and the second (which may be the same) is called its *end-vertex*. The edge $(u \overset{e}{\to} v)$ e is said to be *directed* from u to v. Edges with the same start vertex and the same end vertex are called *parallel*, and if $u \neq v$, $u \overset{e_1}{\to} v$ and $v \overset{e_2}{\to} u$ then e_1 and e_2 are *antiparallel*. An edge $u \to u$ is called a *self-loop*.

The *outdegree*, $d_{out}(v)$, of a vertex v is the number of edges which have v as their start-vertex; *indegree*, $d_{in}(v)$, is defined similarly. Clearly, for every graph

$$\sum_{i=1}^{|v|} d_{in}(v_i) = \sum_{i=1}^{|v|} d_{out}(v_i).$$

A *directed path* is a sequence of edges e_1, e_2, $\ldots$ such that the end vertex of e_{i-1} is the start vertex of e_i. A directed path is a *directed circuit* if the start vertex of the path is the same as its end vertex. The notion of a directed path or circuit being *simple* is defined similarly to that in the undirected case. A digraph is said to be *strongly connected* if for every vertex u and every vertex v there is a directed path from u to v; namely, its *start-vertex* is u and its *end-vertex* is v.

1.2 COMPUTER REPRESENTATION OF GRAPHS

In order to understand the time and space complexities of graph algorithms one needs to know how graphs are represented in the computer memory. In this section two of the most common methods of graph representation are briefly described.

Graphs and digraphs which have no parallel edges are called *simple*. In cases of simple graphs, the specification of the two endpoints is sufficient to specify the edge; in cases of digraph the specification of the start-vertex and end-vertex is sufficient. Thus, we can represent a graph or digraph of n vertices by an $n \times n$ matrix C, where $C_{ij} = 1$ if there is an edge connecting vertex v_i to v_j and $C_{ij} = 0$, if not. Clearly, in the case of graphs $C_{ij} = 1$ implies $C_{ji} = 1$; or in other words, C is symmetric. But in the case of digraphs, any $n \times n$ matrix of zeros and ones is possible. This matrix is called the *adjacency matrix*.

Given the adjacency matrix of a graph, one can compute $d(v_i)$ by counting the number of ones in the i-th row, except that a one on the main diagonal contributes two to the count. For a digraph, the number of ones in the i row is equal to $d_{out}(v_i)$ and the number of ones in the i column is equal to $d_{in}(v_i)$.

The adjacency matrix is not an efficient representation of the graph in case the graph is *sparse*; namely, the number of edges is significantly smaller than n^2. In these cases the following representation, which also allows parallel edges, is preferred.

For each of the vertices, the edges incident to it are listed. This *incidence list* may simply be an array or may be a linked list. We may need a table which tells us the location of the list for each vertex and a table which tells us for each edge its two endpoints (or start-vertex and end-vertex, in case of a digraph).

We can now trace a path starting from a vertex, by taking the first edge on its incidence list, look up its other endpoint in the edge table, finding the incidence list of this new vertex etc. This saves the time of scanning the row of the matrix, looking for a one. However, the saving is real only if n is large and the graph is sparse, for instead of using one bit per edge, we now use edge names and auxiliary pointers necessary in our data structure. Clearly, the space required is $O(|E| + |V|)$, i.e., bounded by a constant times $|E| + |V|$. Here we assume that the basic word length of our computer is large enough to encode all edges and vertices. If this assumption is false then the space required is $O((|E| + |V|) \log (|E| + |V|))$*.

In practice, most graphs are sparse. Namely, the ratio $(|E| + |V|)/|V|^2$ tends to zero as the size of the graphs increases. Therefore, we shall prefer the use of incidence lists to that of adjacency matrices.

---

*The base of the log is unimportant (clearly greater than one), since this estimate is only up to a constant multiplier.

The reader can find more about data structures and their uses in graph theoretic algorithms in references 1 and 2.

1.3 EULER GRAPHS

An *Euler path* of a finite undirected graph $G(V, E)$ is a path $e_1, e_2, \ldots, e_l$ such that every edge appears on it exactly once; thus, $l = |E|$. An undirected graph which has an Euler path is called an *Euler graph*.

Theorem 1.1: A finite (undirected) connected finite graph is an Euler graph if and only if exactly two vertices are of odd degree or all vertices are of even degree. In the latter case, every Euler path of the graph is a circuit, and in the former case, none is.

As an immediate conclusion of Theorem 1.1 we observe that none of the graphs in Figure 1.3 is an Euler graph, because both have four vertices of odd degree. The graph shown in Figure 1.3(a) is the famous *Königsberg bridge problem* solved by Euler in 1736. The graph shown in Figure 1.3(b) is a common misleading puzzle of the type "draw without lifting your pen from the paper".

Proof: It is clear that if a graph has an Euler path which is not a circuit, then the start vertex and the end vertex of the path are of odd degree, while all the other vertices are of even degree. Also, if a graph has a Euler circuit, then all vertices are of even degree.

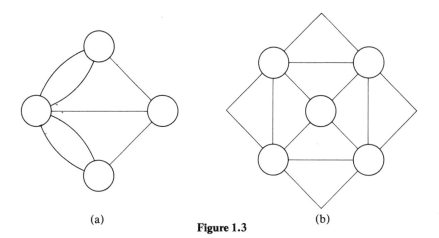

(a)   (b)

Figure 1.3

Assume now that G is a finite graph with exactly two vertices of odd degree, a and b. We shall described now an algorithm for finding a Euler path from a to b. Starting from a we choose any edge adjacent to it (an edge of which a is an endpoint) and trace it (go to its other endpoint). Upon entering a vertex we search for an unused incident edge. If the vertex is neither a nor b, each time we pass through it we use up two of its incident edges. The degree of the vertex is even. Thus, the number of unused incident edges after leaving it is even. (Here again, a self-loop is counted twice.) Therefore, upon entering it there is at least one unused incident edge to leave by. Also, by a similar argument, whenever we reenter a we have an unused edge to leave by. It follows that the only place this process can stop is in b. So far we have found a path which starts in a, ends in b, and the number of unused edges incident to any vertex is even. Since the graph is connected, there must be at least one unused edge which is incident to one of the vertices on the existing path from a to b. Starting a trail from this vertex on unused edges, the only vertex in which this process can end (because no continuation can be found) is the vertex in which it started. Thus, we have found a circuit of edges which were not used before, and in which each edge is used at most once: it starts and ends in a vertex visited in the previous path. It is easy to change our path from a to b to include this detour. We continue to add such detours to our path as long as not all edges are in it.

The case of all vertices of even degrees is similar. The only difference is that we start the initial tour at any vertex, and this tour must stop at the same vertex. This initial circuit is amended as before, until all edges are included.

Q.E.D.

In the case of digraphs, a *directed Euler path* is a directed path in which every edge appears exactly once. A *directed Euler circuit* is defined similarly. Also a *digraph* is called *Euler* if it has a directed Euler path (or circuit).

The *underlying* (undirected) *graph* of a digraph is the graph resulting from the digraph if the direction of the edges is ignored. Thus, the underlying graph of the digraph shown in Figure 1.4(a) is shown in Figure 1.4(b).

Theorem 1.2: A finite digraph is an Euler digraph if any only if its underlying graph is connected and one of the following two conditions holds:

1. There is one vertex a such that $d_{out}(a) = d_{in}(a) + 1$ and another vertex b such that $d_{out}(b) + 1 = d_{in}(b)$, while for all other vertices v, $d_{out}(v) = d_{in}(v)$.
2. For all vertices v, $d_{out}(v) = d_{in}(v)$.

If 1 holds then every directed Euler path starts in a and ends in b. If 2 holds then every directed Euler path is a directed Euler circuit.

(a)   (b)

Figure 1.4

The proof of the theorem is along the same lines as the proof of Theorem 1.1, and will not be repeated here.

Let us make now a few comments about the complexity of the algorithm for finding an Euler path, as described in the proof of Theorem 1.1. Our purposed is to show that the time complexity of the algorithm is $O(|E|)$; namely, there exists a constant K such that the time it takes to find an Euler path is bounded by $K \cdot |E|$.

In the implementation, we use the following data structures:

1. Incidence lists which describe the graph.
2. A doubly-linked list P describing the path. Initially this list is empty.
3. A vertex table, specifying for each vertex v the following data:
    (a) A mark which tells whether v appears already on the path. Initially all vertices are marked "unvisited".
    (b) A pointer $N(v)$, to the next edge on the incidence list, which is the first not to have been traced from v before. Initially $N(v)$ points to the first edge on v's incidence list.
    (c) A pointer $E(v)$ to an edge on the path which has been traced from v. Initially $E(v)$ is "undefined".
4. An edge table which specified for each edge its two endpoints and whether it has been used. Initially, all edges are marked "unused".
5. A list L of vertices all of which have been visited. Each vertex enters this list at most once.

First let us describe a subroutine TRACE(d, P), where d is a vertex and P is a doubly linked list, initially empty, for storage of a traced path. The tracing starts in d and ends when the path, stored in P, cannot be extended.

TRACE(d, P):
  (1) $v \leftarrow d$
  (2) If v is "unvisited", put it in L and mark it "visited".
  (3) If $N(v)$ is "used" but is not last on v's incidence list then have $N(v)$ point to the next edge and repeat (3).
  (4) If $N(v)$ is "used" and it is the last edge on v's incidence list then stop.

(5) $e \leftarrow N(v)$

(6) Add e to the end of P.

(7) If $E(v)$ is "undefined" then $E(v)$ is made to point to the occurrence of e in P.

(8) Mark e "used".

(9) Use the edge table to find the other endpoint u of e.

(10) $v \leftarrow u$ and go to (2).

The algorithm is now as follows:

(1) $d \leftarrow a$

(2) TRACE(d, P). [Comment: The subroutine finds a path from a to b.]

(3) If L is empty, stop.

(4) Let u be in L. Remove u from L.

(5) Start a new doubly linked list of edges, P', which is initially empty. [Comment: P' is to contain the detour from u.]

(6) TRACE(u, P')

(7) Incorporate P' into P at $E(u)$. [Comment: This joins the path and the detour into one, possibly longer path. (The detour may be empty.) Since the edge $E(u)$ starts from u, the detour is incorporated in a correct place.]

(8) Go to (3).

It is not hard to see that both the time and space complexity of this algorithm is $O(|E|)$.

1.4   DE BRUIJN SEQUENCES

Let $\Sigma = \{0, 1, \ldots, \sigma - 1\}$ be an alphabet of σ letters. Clearly there are σ^n different words of length n over Σ. A *de Bruijn sequence** is a (circular) sequence $a_0 a_1 \cdots a_{L-1}$ over Σ such that for every word w of length n over Σ there exists a unique i such that

$$a_i a_{i+1} \cdots a_{i+n-1} = w,$$

where the computation of the indices is modulo L. Clearly if the sequence satisfies this condition, the $L = \sigma^n$. The most important case is that of $\sigma = 2$.

---

*Sometimes they are called *maximum-length shift register sequences*.

Binary de Bruijn sequences are of great importance in coding theory and are implemented by shift registers. (See Golomb's book [3] on the subject.) The interested reader can find more information on de Bruijn sequences in references 4 and 5. The only problem we shall discuss here is the existence of de Bruijn sequences for every $\sigma \geq 2$ and every n.

Let us describe a digraph $G_{\sigma,n}(V, E)$ which has the following structure:

1. V is the set of all σ^{n-1} words of length $n - 1$ over Σ.
2. E is the set of all σ^n words of length n over Σ.
3. The edge $b_1 b_2 \cdots b_n$ starts at vertex $b_1 b_2 \cdots b_{n-1}$ and ends at vertex $b_2 b_3 \cdots b_n$.

The graphs $G_{2,3}$, $G_{2,4}$, and $G_{3,2}$ are shown in Figures 1.5, 1.6 and 1.7 respectively.

These graphs are sometimes called de Bruijn diagrams, or Good's diagrams, or shift register state diagrams. The structure of the graphs is such that the word w_2 can follow the word w_1 in a de Bruijn sequence only if the edge w_2 starts at the vertex in which w_1 ends. Also it is clear that if we find a directed Euler circuit (a directed circuit which uses each of the graph's edges exactly once) of $G_{\sigma,n}$, then we also have a de Bruijn sequence. For example, consider the directed Euler circuit of $G_{2,3}$ (Figure 1.5) consisting of the following sequence of edges:

$$000, \ 001, \ 011, \ 111, \ 110, \ 101, \ 010, \ 100.$$

The implied de Bruijn sequence, 00011101, follows by reading the first letter of each word in the circuit. Thus, the question of existence of de Bruijn sequences is equivalent to that of the existence of direct Euler circuits in the corresponding de Bruijn diagram.

Theorem 1.3: For every positive integers σ and n, $G_{\sigma,n}$ has a directed Euler circuit.

Proof: We wish to use Theorem 1.2 to prove our theorem. First we have to show that the underlying undirected graph is connected. In fact, we shall show that $G_{\sigma,n}$ is strongly connected. Let $b_1 b_2 \cdots b_{n-1}$ and $c_1 c_2 \cdots c_{n-1}$ be any two vertices; the directed path $b_1 b_2 \cdots b_{n-1} c_1$, $b_2 b_3 \cdots b_{n-1} c_1 c_2$, ..., $b_{n-1} c_1 c_2 \cdots c_{n-1}$ leads from the first to the second. Next, we have to show that $d_0(v) = d_1(v)$ for each vertex v. The vertex $b_1 b_2 \cdots b_{n-1}$ is entered by

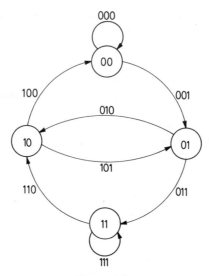

Figure 1.5

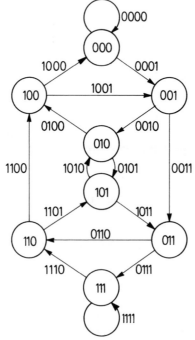

Figure 1.6

edges $cb_1b_2 \cdots b_{n-1}$, where c can be chosen in σ ways, and is the start vertex of edges $b_1b_2 \cdots b_{n-1}c$, where again c can be chosen in σ ways.

Q.E.D.

Corollary 1.1: For every positive integers σ and n there exists a de Bruijn sequence:

1.5 SHORTEST-PATH ALGORITHMS

In general the shortest-path problems are concerned with finding shortest paths between vertices. Many interesting problems arise, and the variety depends on the type of graph in our application and the exact question we want to answer. Some of the characteristics which may help in defining the exact problem are as follows:

1. The graph is finite or infinite.
2. The graph is undirected or directed.
3. The edges are all of length 1, or all lengths are non-negative, or negative lengths are allowed.
4. We may be interested in shortest paths from a given vertex to another, or from a given vertex to all the other vertices, or from each vertex to all the other vertices.
5. We may be interested in finding just one path, or all paths, or counting the number of shortest paths.

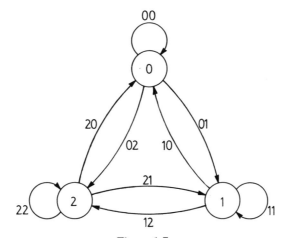

Figure 1.7

Clearly, this section will deal only with very few of all the possible problems. An attempt is made to describe the most important techniques.

First let us consider the case of a finite graph G in which two vertices s and t are specified. Our task is to find a path from s to t, if there are any, which uses the least number of edges. Clearly this is the case of the finite, undirected graph, with all length of edges being equal to 1, and where all we want is one path from a given vertex to another. In fact, the digraph case is just as easy and can be similarly solved.

The algorithm to be used here was suggested by Moore [6] and by now is widely used. It is well known as the *Breadth First Search (BFS)* technique.

At first no vertices of the graph are considered labeled.

1. Label vertex s with 0.
2. $i \leftarrow 0$
3. Find all unlabeled vertices adjacent to at least one vertex labeled i. If none are found, stop.
4. Label all the vertices found in (3) with $i + 1$.
5. If vertex t is labeled, stop.
6. $i \leftarrow i + 1$ and go to (3).

Clearly we can remove step 5 from the algorithm, and the algorithm is still valid for finite graphs. However, step 5 saves the work which would be wasted after t is labeled, and it permits the use of the algorithm on infinite graphs whose vertices are of finite degree and in which there is a (finite) path between s and t.

Let the *distance* between u and v be the least number of edges in a path connecting u and v, if such a path exists, and ∞ if none exists.

Theorem 1.3: The BFS algorithm computes the distance of each vertex from s, if t is not closer.

Proof: Let us denote the label of a vertex v, assigned by the BFS algorithm, by $\lambda(v)$.

First we show that if a vertex is labeled $\lambda(v) = k$, then there is a path of length k from s to v. Such a path can be traced as follows: There must be a vertex v_{k-1} adjacent to $v = v_k$, labeled $k - 1$, and similarly, there must be a vertex v_{k-i-1} adjacent to v_{k-i} labeled $k - i - 1$ for $i = 0, 1, \ldots, k - 1$. Clearly $v_0 = s$, since s is the only vertex labeled 0. Thus, $v_0 - v_1 - \cdots v_{k-1} - v_k$ is a path of length k from s to v.

Now, let us prove by induction on l that if v is of distance l from s and if t is not closer to s, then $\lambda(v) = l$.

After Step 1, $\lambda(s) = 0$, and indeed the distance from s to s is zero.

Assume now that the statement holds for shorter distances, let us show that it must hold for l too. Let $s - v_1 - v_2 - \cdots v_{l-1} - v$ be a shortest path from s to v. Clearly, $s - v_1 - v_2 \cdots v_{l-2} - v_{l-1}$ is a shortest path from s to v_{l-1}. If t is not closer to s than v then clearly it is not closer than v_{l-1} either. By the inductive hypothesis $\lambda(v_{l-1}) = l - 1$. When $i = l - 1$, v receives the label l. It could not have been labeled before since if it were then its label is less than l, and there is a shorter path from s to v, in contradiction to l's definition.

Q.E.D.

It is clear that each edge is traced at most twice, in this algorithm; once from each of its endpoints. That is, for each i the vertices labeled i are scanned for their incident edges in step 3. Thus, if we use the incidence lists data structures the algorithm will be of time complexity $O(|E|)$.

The directed case in even simpler because each edge is traced at most once.

A path from s to t can be traced by moving now from t to s, as described in the proof of Theorem 1.3. If we leave for each vertex the name of the edge used for labeling it, the tracing is even easier.

Let us now consider the case of a finite digraph, whose edges are assigned with non-negative length; thus, each edge e is assigned a length $l(e) \geq 0$. Also, there are two vertices s and t and we want to find a shortest directed path from s to t, where the length of a path is the sum of the lengths of its edges.

The following algorithm is due to Dijkstra [7]:

1. $\lambda(s) \leftarrow 0$ and for all $v \neq s$, $\lambda(v) \leftarrow \infty$.
2. $T \leftarrow V$.
3. Let u be a vertex in T for which $\lambda(u)$ is minimum.
4. If $u = t$, stop.
5. For every edge $u \xrightarrow{e} v$, if $v \in T$ and $\lambda(v) > \lambda(u) + l(e)$ then $\lambda(v) \leftarrow \lambda(u) + l(e)$.
6. $T \leftarrow T - \{u\}$ and go to step 3.

Let us denote the distance of vertex v from s by $\delta(v)$. We want to show that upon termination $\delta(t) = \lambda(t)$; that is, if $\lambda(t)$ is finite than it is equal to $\delta(t)$ and if $\lambda(t)$ is infinite then there is no path from s to t in the digraph.

Lemma 1.2: In Dijkstra's algorithm, if $\lambda(v)$ is finite then there is a path from s to v whose length is $\lambda(v)$.

Proof: Let u be the vertex which gave v its present label $\lambda(v)$; namely, $\lambda(u) + l(e) = \lambda(v)$, where $u \overset{e}{\to} v$. After this assignment took place, u did not change its label, since in the following step (step 6) u was removed from the set T (of temporarily assigned vertices) and its label remained fixed from there on. Next, find the vertex which gave u its final label $\lambda(u)$, and repeating this backward search, we trace a path from s to v whose length is exactly $\lambda(v)$. The backward search finds each time a vertex which has left T earlier, and therefore no vertex on this path can be repeated; it can only terminate in s which has been assigned its label in step 1.

Q.E.D.

Lemma 1.3: In Dijkstra's algorithm, when a vertex is chosen (in Step 3), its label $\lambda(u)$ satisfies $\lambda(u) = \delta(u)$.

Proof: By induction on the order in which vertices leave the set T. The first one to leave is s, and indeed $\lambda(s) = \delta(s) = 0$.

Assume now that the statement holds for all vertices which left T before u.

If $\lambda(u) = \infty$, let u' be the first vertex whose label $\lambda(u')$ is infinite when it is chosen. Clearly, for every v in T, at this point, $\lambda(v) = \infty$, and for all vertices $v' \in V - T$, $\lambda(v')$ is finite. Therefore, there is no edge with a start-vertex in $V - T$ and end-vertex in T. It follows that there is no path from s (which is in $V - T$) to u (which is in T).

If $\lambda(u)$ is finite, then by Lemma 1.2, $\lambda(u)$ is the length of some path from s to u. Thus, $\lambda(u) \geq \delta(u)$. We have to show that $\lambda(u) > \delta(u)$ is impossible. Let a shortest path from s to u be $s = v_0 \overset{e_1}{\to} v_1 \overset{e_2}{\to} \cdots v_{k-1} \overset{e_k}{\to} v_k = u$. Thus, for every $i = 0, 1, \ldots, k$

$$\delta(v_i) = \sum_{j=1}^{i} l(e_j).$$

Let v_i be the right most vertex on this path to leave T before u. By the inductive hypothesis

$$\lambda(v_i) = \delta(v_i) = \sum_{j=1}^{i} l(e_j).$$

If $v_{i+1} \neq u$, then $\lambda(v_{i+1}) \leq \lambda(v_i) + l(e_{i+1})$ after v_i has left T. Since labels can only decrease if they change at all, when u is chosen $\lambda(v_{i+1})$ still satisfies this inequality. We have:

$$\lambda(v_{i+1}) \leq \lambda(v_i) + l(e_{i+1}) = \delta(v_i) + l(e_{i+1}) = \delta(v_{i+1}) \leq \delta(u),$$

and if $\delta(u) < \lambda(u)$, u should not have been chosen. In case $v_{i+1} = u$, the same argument shows directly that $\lambda(u) \leq \delta(u)$.

Q.E.D.

It is an immediate corollary of Lemma 1.3 that $\lambda(t) = \delta(t)$ upon termination.

Let us now consider the complexity of Dijkstra's algorithm. In step 3, the minimum label of the elements of T has to be found. Clearly this can be done in $|T| - 1$ comparisons. At first $T = V$; it decreases by one each time and the search is repeated $|V|$ times. Thus, the total time spent on step 3 is $O(|V|^2)$. Step 5 can use each edge exactly once. Thus it uses, at most, $O(|E|)$ time. Since it makes no sense to have parallel edges (for all but the shortest can be dropped) or self-loops, $|E| \leq |V| \cdot (|V| - 1)$. Thus, the whole algorithm is of $O(|V|^2)$ complexity.

Clearly, for sparse graphs the BFS algorithm is better; unfortunately it does not work if not all edge lengths are equal.

Dijkstra's algorithm is applicable to undirected graphs too. Simply represent each edge of the graph by two anti-parallel directed edges with the same length. Also, it can be used on infinite graphs, if outgoing degrees are finite and there is a finite directed path from s to t. However, this algorithm is not applicable if $l(e)$ may be negative; Lemma 1.2 still holds, but Lemma 1.3 does not.

Next, an algorithm for finding the distance of all the vertices of a finite digraph from a given vertex s, is described. It allows negative edge lengths, but does not allow a directed circuit whose length (sum of the lengths of its edges) is negative. The algorithm is due to Ford [8, 9]:

1. $\lambda(s) \leftarrow 0$ and for every $v \neq s$, $\lambda(v) \leftarrow \infty$.
2. As long as there is an edge $u \xrightarrow{e} v$ such that $\lambda(v) > \lambda(u) + l(e)$ replace $\lambda(v)$ by $\lambda(u) + l(e)$.

For our purposes ∞ is not greater than $\infty + k$, even if k is negative.

It is not even obvious that the algorithm will terminate; indeed, it will not if there is a directed circuit accessible from s (namely, there is a directed path from s to one of the vertices on the circuit) whose length is negative. By going around this circuit the labels will be decreased and the process can be repeated indefinitely.

Lemma 1.4: In the Ford algorithm, if $\lambda(v)$ is finite then there is a directed path from s to v whose length is $\lambda(v)$.

Proof: As in the proof of the similar previous statements, this is proved by displaying a path from s to v, and its construction is backwards, from v to s. First we find the vertex u which gave v its present label $\lambda(v)$. The value of $\lambda(u)$ may have decreased since, but we shall refer to its value $\lambda'(u)$ at the time that it gave v its label. Thus, $\lambda(v) = \lambda'(u) + l(e)$, where $u \overset{e}{\to} v$. We continue from u to the vertex which gave it the value $\lambda'(u)$ etc., each time referring to an earlier time in the running of the algorithm. Therefore, this process must end, and the only place it can end is in s.

Q.E.D.

The lemma above is even true if there are negative length directed circuits. But if there are no such circuits, the path traced in the proof cannot return to a vertex visited earlier. For if it does, then by going around the directed circuit, a vertex improved its own label; this implies that the sum of the edge lengths of the circuit is negative. Therefore we have:

Lemma 1.5: In the Ford algorithm, if the digraph has no directed circuits of negative length and if $\lambda(v)$ is finite then there is a simple directed path from s to v whose length is $\lambda(v)$.

Since each value, $\lambda(v)$, corresponds to at least one simple path from s to v, and since the number of simple directed paths in a finite digraph is finite, the number of values possible for $\lambda(v)$ is finite. Thus, the Ford algorithm must terminate.

Lemma 1.6: For a digraph with no negative directed circuit, upon termination of the Ford algorithm, $\lambda(v) = \delta(v)$ for every vertex v.

Proof: By Lemma 1.5, $\lambda(v) \geq \delta(v)$. If $\lambda(v) > \delta(v)$, let $s = v_0 \overset{e_1}{\to} v_1 \to \cdots$ $v_{k-1} \overset{e_k}{\to} v_k = v$ be a shortest path from s to v. Clearly, for every $i = 0, 1, \ldots, k$

$$\delta(v_i) = \sum_{j=1}^{i} l(e_j).$$

Let v_i be the first vertex on this path for which $\lambda(v_i) > \delta(v_i)$. Since $\lambda(v_{i-1}) = \delta(v_{i-1})$, the edge $v_{i-1} \overset{e_i}{\to} v_i$ can be used to lower $\lambda(v_i)$ to $\lambda(v_{i-1}) + l(e_i)$, (which is equal to $\delta(v_i)$). Thus, the algorithm should not have terminated.

Q.E.D.

We can use a simple device to bound the number of operation to $O(|E| \cdot |V|)$. Order the edges: $e_1, e_2, \ldots, e_{|E|}$. Now, perform step 2 by first checking e_1, then e_2, etc., and improving labels accordingly. After the first such sweep, go through additional sweeps, until an entire sweep produces no improvement. If the digraph contains no negative directed circuits then the process will terminate. Furthermore, if a shortest path from s to v consists of k edges, then by the end of the kth sweep v will have its final label; this is easily proved by induction on k. Since k is bounded by $|V|$, the whole algorithm takes at most $O(|E| \cdot |V|)$ steps. Moreover, if by the $|V|$th sweep any improvement of a label takes place then the digraph must contain a negative circuit. Thus, we can use the Ford algorithm to detect the existance of a negative circuit, if all vertices are accessible from s. If the existence of a negative circuit is indicated, we can find one by starting a backward search from a vertex whose label is improved in the $|V|$th sweep. To this end we need for each vertex v a pointer to the vertex u which gave v its last label. This is easily achieved by a simple addition to step 2.

The Ford algorithm cannot be used on undirected graphs because any negative edge has the effect of a negative circuit; one can go on it back and forth decreasing labels indefinitely. All three algorithms can be used to find the distances of all vertices from a given vertex s; the BFS and Dijkstra's algorithm have to be changed: instead of stoping when t is labeled or taken out of T, stop when all accessible vertices are labeled or when T is empty. The bounds on the number of operations remain $O(|E|)$, $O(|V|^2)$ and $O(|E| \cdot |V|)$ respectively for the BFS, Dijkstra and the Ford algorithm. If this is repeated from all vertices, in order to find the distance from every vertex to all the others, the respective complexities are $O(|E| \cdot |V|)$, $O(|V|^3)$ and $O(|E| \cdot |V|^2)$. Next, let us describe an algorithm which solves the case with negative lengths in time $O(|V|^3)$.

Let $G(V, E)$ be a finite digraph with $V = \{1, 2, \ldots, n\}$. The length of edge e is denoted by $l(e)$, as before, and it may be negative. Define

$$\delta^0(i, j) = \begin{cases} l(e) & \text{if } i \xrightarrow{e} j, \\ \infty & \text{if there is no edge from } i \text{ to } j. \end{cases}$$

Let $\delta^k(i, j)$ be the length of a shortest path from i to j among all paths which may pass through vertices $1, 2, \ldots, k$ but do not pass through vertices $k + 1, k + 2, \ldots, n$.

Floyd's algorithm [10] is as follows:

1. $k \leftarrow 1$
2. For every $1 \leq i, j \leq n$ compute

$$\delta^k(i, j) \leftarrow Min\{\delta^{k-1}(i, j), \delta^{k-1}(i, k) + \delta^{k-1}(k, j)\}.$$

3. If $k = n$, stop. If not, increment k and go to step 2.

The algorithm clearly yields the right answer; namely, $\delta^n(i, j)$ is the distance from i to j. The answer is only meaningful if there are no negative circuits in G. The existence of negative circuits is easily detected by $\delta^k(i, i) < 0$. Each application of step 2 requires n^2 operations, and step 2 is repeated n times. Thus, the algorithm is of complexity $O(n^3)$.

For the case of finite graphs with non-negative edge lengths, both the repeated Dijkstra algorithm and Floyd's take $O(|V|^3)$. Additional information on the shortest path problem can be found in Problems 1.9 and 1.10 and references 11 and 12.

PROBLEMS

1.1   Prove that if a connected (undirected) finite graph has exactly $2k$ vertices of odd degree then the set of edges can be partitioned into k paths such that every edge is used exactly once. Is the condition of connectivity necessary or can it be replaced by a weaker condition?

A *Hamilton path* (circuit) is a simple path (circuit) on which every vertex of the graph appears exactly once.

1.2   Prove that the following graph has no Hamilton path or circuit.

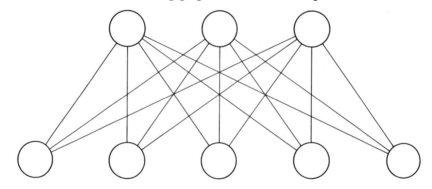

1.3 Prove that in every completely connected directed graph (a graph in which every two vertices are connected by exactly one directed edge in one of the two possible directions) there is always a directed Hamilton path. (Hint: Prove by induction on the number of vertices.)

1.4 Prove that a directed Hamilton circuit of $G_{o,n}$ corresponds to a directed Euler circuit of $G_{o,n-1}$. It is true that $G_{o,n}$ always has a direct Hamilton circuit?

1.5 In the following assume that $G(V, E)$ is a finite undirected graph, with no parallel edges and no self-loops.

(a) Describe an algorithm which attempts to find a Hamilton circuit in G by working with a partial simple path. If the path cannot be extended in either direction then try to close it into a simple circuit by the edge between its endpoints, if it exists, or by a switch, as suggested by the diagram, where edges a and b are added and c is deleted. Once a circuit is formed, look for an edge from one of its vertices to a new vertex, and open the circuit to a now longer path, etc.

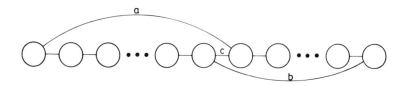

(b) Prove that if for every two vertices u and v, $d(u) + d(v) \geq n$, where $n = |V|$, then the algorithm will never fail to produce a Hamilton circuit.

(c) Deduce Dirac's theorem [13]: If for every vertex v, $d(v) \geq n/2$, then G has a Hamilton circuit.

1.6 Describe an algorithm for finding the number of shortest paths from s to t after the BFS algorithm has been performed.

1.7 Repeat the above, after the Dijkstra algorithm has been performed. Assume $l(e) > 0$ for every edge e. Why is this asumption necessary?

1.8 Prove that a connected undirected graph G is orientable (by giving each edge some direction) into a strongly connected digraph if and only if each edge of G is in some simple circuit in G. (A path $u \xrightarrow{e} v \xrightarrow{e} u$ is not considered a simple circuit.)

1.9 The *transitive closure* of a digraph $G(V, E)$ is a digraph $G'(V, E)$ such that there is an edge $u \to v$ in G' if and only if there is a (non-empty) directed path from u to v in G. For the BFS, Dijkstra and Floyd's algorithms, explain how they can be used to construct G' for a given G, and compare the complexities of the resulting algorithms.

1.10 The following algorithm, due to Dantzig [14], finds all distances in a finite digraph, like Floyd's algorithm. Let $\delta^k(i, j)$ be the distance from i to j, where $1 \le i, j \le k$ and no vertices higher than k are used on the path. Let $\delta^k(i, i) = 0$ for all i and k. Also, let $l(i, j)$ be $l(e)$ if $i \overset{e}{\to} j$, and ∞ if no such edge exists.

(1) $\delta^1(1, 1) \leftarrow$ Min $\{0, l(1, 1)\}$.

(2) $k \leftarrow 2$

(3) For $1 \le i < k$ do
$$\delta^k(i, k) \leftarrow \text{Min}_{1 \le j < k}\{\delta^{k-1}(i, j) + l(j, k)\}$$
$$\delta^k(k, i) \leftarrow \text{Min}_{1 \le j < k}\{l(k, j) + \delta^{k-1}(j, i)\}$$

(4) For $1 \le i, j < k$ do
$$\delta^k(i, j) \leftarrow \text{Min}\{\delta^{k-1}(i, j), \delta^k(i, k) + \delta^k(k, j)\}.$$

(5) If $k = n$, stop, If not, increment k and go to step 3.

Show that Dantzig's algorithm is valid. How are negative circuits detected? What is the time complexity of this algorithm?

REFERENCES

[1] Knuth, D. E., *The Art of Computer Programming, Vol. 1: Fundamental Algorithms*, Addison-Wesley, 1968.

[2] Aho, A. V., Hopcroft, J. E., and Ullman, J. D., *The Design and Analysis of Computer Algorithms*, Addison-Wesley, 1974.

[3] Golomb, S. W., *Shift Register Sequences*, Holden-Day, 1967.

[4] Berge, C., *The Theory of Graphs and Its Applications*, Wiley, 1962, Chapter 17.

[5] Hall, M., Jr., *Combinatorial Theory*, Blaisdell, 1967, Chapter 9.

[6] Moore, E. F., "The Shortest Path Through a Maze", *Proc. Iternat. Symp. Switching Th.*, 1957, Part II, Harvard Univ. Press, 1959, pp. 285-292.

[7] Dijkstra, E. W., "A Note on Two Problems in Connection with Graphs", *Numerische Math.*, Vol. 1, 1959, pp. 269-271.

[8] Ford, L. R., Jr., "Network Flow Theory", The Rand Corp., P-923, August, 1956.

[9] Ford, L. R., Jr. and Fulkerson, D. R., *Flows in Networks*, Princeton Univ. Press, 1962, Chap. III, Sec. 5.

[10] Floyd, R. W., "Algorithm 97: Shortest Path", *Comm. ACM*, Vol. 5, 1962, p. 345.

[11] Dreyfus, S. E., "An Appraisal of Some Shortest-Path Algorithms", *Operations Research*, Vol. 17, 1969, pp. 395–412.

[12] Lawler, E. L., *Combinatorial Optimization: Networks and Matroids*, Holt, Rinehart and Winston, 1976, Chapter 3.

[13] Dirac, G. A., "Connectivity Theorems for Graphs", *Quart. J. Math.*, Ser. (2), Vol. 3, 1952, pp. 171–174.

[14] Dantzig, G. B., "All Shortest Routes in a Graph", Oper. Res. House, Stanford Univ. Tech. Rep. 66-3, November 1966.

Chapter 2

TREES

2.1 TREE DEFINITIONS

Let $G(V, E)$ be an (undirected), finite or infinite graph. We say that G is *circuit-free* if there are no simple circuits in G. G is called a *tree* if it is connected and circuit-free.

Theorem 2.1: The following four conditions are equivalent:

(a) G is a tree.
(b) G is circuit-free, but if any new edge is added to G, a circuit is formed.
(c) G contains no self-loops and for every two vertices there is a unique simple path connecting them.
(d) G is connected, but if any edge is deleted from G, the connectivity of G is interrupted.

Proof: We shall prove that conditions (a) $\Rightarrow$ (b) $\Rightarrow$ (c) $\Rightarrow$ (d) $\Rightarrow$ (a).

(a) $\Rightarrow$ (b): We assume that G is connected and circuit-free. Let e be a new edge, that is $e \notin E$; the two endpoints of e, a and b, are elements of V. If $a = b$, then e forms a self-loop and therefore a circuit exists. If $a \neq b$, there is a path in G (without e) between a and b; if we add e, this path with e forms a circuit.

(b) $\Rightarrow$ (c): We assume that G is circuit-free and that no edge can be added to G without creating a circuit. Let a and b be any two vertices of G. If there is no path between them, then we can add an edge between a and b without creating a circuit. Thus, G must be connected. Moreover, if there are two simple paths, P and P', between a and b, then there is a circuit in G. To see this, assume that $P = e_1, e_2, \ldots, e_l$ and $P' = e_1', e_2', \ldots, e_m'$. Since both paths are simple, one cannot be the beginning of the other. Let i be the first index for which $e_i \neq e_i'$, and let v be the first vertex on $e_i, e_{i+1}, \ldots, e_l$ which is also on $e_i', e_{i+1}', \ldots, e_m'$. The two

22

disjoint subpaths between the branching off vertex and v form a simple circuit in G.

(c) $\Rightarrow$ (d): We assume the existence of a unique simple path between every pair of vertices of G. This implies that G is connected. Assume now that we delete an edge e from G. Since G has no self-loops, e is not a self-loop. Let a and b be e's endpoints. If there is now (after the deletion of e) a path between a and b, then G has more than one simple path between a and b.

(d) $\Rightarrow$ (a): We assume that G is connected and that no edge can be deleted without interrupting the connectivity. If G contains a simple circuit, any edge on this circuit can be deleted without interrupting the connectivity. Thus, G is circuit-free.

Q.E.D.

There are two more common ways to define a finite tree. These are given in the following theorem.

Theorem 2.2: Let $G(V, E)$ be a finite graph and $n = |V|$. The following three conditions are equivalent:

(a) G is a tree.
(b) G is circuit-free and has $n - 1$ edges.
(c) G is connected and has $n - 1$ edges.

Proof: For $n = 1$ the theorem is trivial. Assume $n \geq 2$. We shall prove that conditions (a) $\Rightarrow$ (b) $\Rightarrow$ (c) $\Rightarrow$ (a).

(a) $\Rightarrow$ (b): Let us prove, by induction on n, that if G is a tree, then its number of edges is $n - 1$. This statement is clearly true for $n = 1$. Assume that it is true for all $n < m$, and let G be a tree with m vertices. Let us delete from G any edge e. By condition (d) of Theorem 2.1, G is not connected any more, and clearly is broken into two connected components each of which is circuit-free and therefore is a tree. By the inductive hypothesis, each component has one edge less than the number of vertices. Thus, both have $m - 2$ edges. Add back e, and the number of edges is $m - 1$.

(b) $\Rightarrow$ (c): We assume that G is circuit-free and has $n - 1$ edges. Let us first show that G has at least two vertices of degree 1. Choose any edge e. An edge must exist since the number of edges is $n - 1$ and $n \geq 2$. Extend the edge into a path by adding new edges to its ends if such exist. A new edge attached at the path's end introduces a new vertex to the path or a circuit is closed. Thus, our path remains simple. Since the graph is finite, this extension must terminate on both sides of e, yielding two vertices of degree 1.

Now, the proof that G is connected proceeds by induction on the number of vertices, n. The statement is obviously true for $n = 2$. Assume that it is true for $n = m - 1$, and let G be a circuit-free graph with m vertices and $m - 1$ edges. Eliminate from G a vertex v, of degree 1, and its incident edge. The resulting graph is still circuit-free and has $m - 1$ vertices and $m - 2$ edges; thus, by the inductive hypothesis it is connected. Therefore, G is connected too.

(c) $\Rightarrow$ (a): Assume that G is connected and has $n - 1$ edges. If G contains circuits, we can eliminate edges (without eliminating vertices) and maintain the connectivity. When this process terminates, the resulting graph is a tree, and, by (a) $\Rightarrow$ (b), has $n - 1$ edges. Thus, no edge can be eliminated and G is circuit-free.

Q.E.D.

Let us call a vertex whose degree is 1, a *leaf*. A corollary of Theorem 2.2 and the statement proved in the (b) $\Rightarrow$ (c) part of its proof is the following corollary:

Corollary 2.1: A finite tree, with more than one vertex, has at least two leaves.

2.2 MINIMUM SPANNING TREE

A graph $G'(V', E')$ is called a *subgraph* of a graph $G(V, E)$, if $V' \subseteq V$ and $E' \subseteq E$. Clearly, an arbitrary choice of $V' \subseteq V$ and $E' \subseteq E$ may not yield a subgraph, simply because it may not be a graph; that is, some of the endpoints of edges in E' may not be in V'.

Assume $G(V, E)$ is a finite, connected (undirected) graph and each edge $e \in E$ has a known length $l(e) > 0$. Assume we want to find a connected subgraph $G'(V, E')$ whose length, $\sum_{e \in E} l(e)$, is minimum; or, in other words, we want to remove from G a subset of edges whose total length is maximum, and which leaves it still connected. It is clear that such a subgraph is a tree. For G' is assumed to be connected, and since its length is minimum, none of its edges can be removed without destroying its connectivity. By Theorem 2.1 (see part (d)) G' is a tree. A subgraph of G, which contains all of its vertices and is a tree is called a *spanning tree* of G. Thus, our problem is that of finding a minimum-length spanning tree of G.

There are many known algorithms for the minimum spanning tree problem, but they all hinge on the following theorem:

Theorem 2.3: Let $U \subset V$ and e be of minimum length among the edges with one endpoint in U and the other endpoint in $V - U$. There exists a minimum spanning tree T such that e is in T.

Proof: Let T_0 be a minimum spanning tree. If e is not in T_0, add e to T_0. By Theorem 2.1 (part (b)) a circuit is formed. This circuit contains e and at least one more edge $u \overset{e'}{\text{---}} v$, where $u \in U$ and $v \in V - U$. Now, $l(e) \leq l(e')$, since e is of minimum length among the edges connecting U with $V - U$. We can delete e' from $T_0 + e$. The resulting subgraph is still connected and by Theorem 2.2 is a tree, since it has the right number of edges. Also, the length of this new tree, which contains e, is less than or equal to that of T_0. Thus, it is optimal.

Q.E.D.

Let $G(V, E)$ be the given graph, where $V = \{1, 2, \ldots, n\}$. We assume that there are no parallel edges, for all but the shortest can be eliminated. Thus, let $l(i, j)$ be $l(e)$ if there is an edge $i \overset{e}{\text{---}} j$, and infinity otherwise. The following algorithm is due to Prim [1]:

(1) $t \leftarrow 1$, $T \leftarrow \emptyset$ and $U \leftarrow \{1\}$.
(2) Let $l(t, u) = \text{Min}_{v \in V - U}\{l(t, v)\}$.
(3) $T \leftarrow T \cup \{e\}$ where e is the edge which corresponds to the length $l(t, u)$.
(4) $U \leftarrow U \cup \{u\}$.
(5) If $U = V$, stop.
(6) For every $v \in V - U$, $l(t, v) \leftarrow \text{Min}\{l(t, v), l(u, v)\}$.
(7) Go to Step (2).

(Clearly $t = 1$ throughout. We used t instead of 1 to emphasize that $l(t, v)$ may not be the original $l(1, v)$ after Step (6) has been applied.)

The algorithm follows directly the hint supplied by Theorem 2.3. The "vertex" t represents the subset U of vertices, and for $v \in V - U$ $l(t, v)$ is the length of a shortest edge from a vertex in U to v. This is affected by Step (6). Thus, in Step (2), a shortest edge connecting U and $V - U$ is chosen.

Although each choice of an edge is "plausible", it is still necessary to prove that in the end, T is a minimum spanning tree.

Let a subgraph $G'(V', E')$ be called an *induced subgraph* if E' contains all the edges of E whose endpoints are in V'; in this case we say that G' is induced by V'.

First observe, that each time we reach Step (5), T is the edge set of a spanning tree of the subraph induced by U. This is easily proved by induction on the number of times we reach Step (5). We start with $U = \{1\}$ and $T = \emptyset$ which is clearly a spanning tree of the subgraph induced by $\{1\}$. After the first application of Steps (2), (3) and (4), we have two vertices in U and an edge in T which connects them. Each time we apply Steps (2), (3) and (4) we add an edge from a vertex of the previous U to a new vertex. Thus the new T is connected too. Also, the number of edges is one less than the number of vertices. Thus, by Theorem 2.2 (part (c)), T is a spanning tree.

Now, let us proceed by induction to prove that if the old T is a subgraph of some minimum spanning tree of G then so is the new one. The proof is similar to that of Theorem 2.3. Let T_0 be a minimum spanning tree of G which contains T as a subgraph, and assume e is the next edge chosen in Step (2) to connect between a vertex of U and $V - U$. If e is not in T_0, add it to T_0 to form $T_0 + e$. It contains a circuit in which there is one more edge, e', connecting a vertex of U with a vertex of $V - U$. By Step (2), $l(e) \le l(e')$, and if we delete e' from $T_0 + e$, we get an minimum spanning tree which contains both T, as a subgraph, and e, proving that the new T is a subgraph of some minimum spanning tree. Thus, in the end T is a minimum spanning tree of G.

The complexity of the algorithm is $O(|V|^2)$; Step (2) requires at most $|V| - 1$ comparisons and is repeated $|V| - 1$ times, yielding $O(|V|^2)$. Step (6) requires one comparison for each edge; thus, the total time spent on it is $O(|E|)$.

It is possible to improve the algorithm and the interested reader is advised to read the Cheriton and Tarjan paper [2]. We do not pursue this here because an understanding of advanced data structures is necessary. The faster algorithms do not use any graph theory beyond the level of this section.

The analogous problem for diagraphs, namely, that of finding a subset of the edges E' whose total length is minimum among those for which (V, E') is a strongly connected subgraph, is much harder. In fact, even the case where $l(e) = 1$ for all edges is hard. This will be discussed in Chapter 10.

2.3 CAYLEY'S THEOREM

In a later section we shall consider the question of the number of spanning trees in a given graph. Here we consider the more restricted, and yet interesting problem, of the number of trees one can define on a given set of vertices, $V = \{1, 2, \ldots, n\}$.

For $n = 3$, there are 3 possible trees, as shown in Figure 2.1. Clearly, for $n = 2$ there is only one tree. The reader can verify, by exhausting all the cases, that for $n = 4$ the number of trees is 16. The following theorem is due to Cayley [3]:

Theorem 2.4: The number of spanning trees for n distinct vertices is n^{n-2}.

The proof to be presented is due to Prüfer [4]. (For a survey of various proofs see Moon [5].)

Proof: Assume $V = \{1, 2, \ldots, n\}$. Let us display a one-to-one correspondence between the set of the spanning trees and the n^{n-2} words of length $n - 2$ over the alphabet $\{1, 2, \ldots, n\}$. The algorithm for finding the word which corresponds to a given tree is as follows:

(1) $i \leftarrow 1$.
(2) Among all leaves of the current tree let j be the least one (i.e., its name is the least integer). Eliminate j and its incident edge e from the tree. The ith letter of the word is the other endpoint of e.
(3) If $i = n - 2$, stop.
(4) Increment i and go to step 2.

For example, assume that $n = 6$ and the tree is as shown in Figure 2.2. On the first turn of Step (2), $j = 2$ and the other endpoint of its incident edge is 4. Thus, 4 is the first letter of the word. The new tree is as shown in Figure 2.3. On the second turn, $j = 3$ and the second letter is 1. On the third, $j = 1$ and the third letter is 6. On the fourth, $j = 5$ and the fourth letter is 4. Now $i = 4$ and the algorithm halts. The resulting word is 4164 (and the current tree consists of one edge connecting 4 and 6).

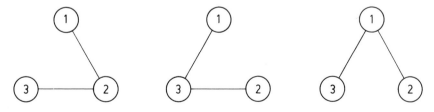

Figure 2.1

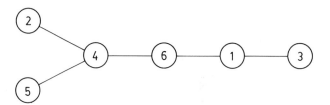

Figure 2.2

Figure 2.3

By Corollary 2.1, Step (2) can always be performed, and therefore for every tree a word of length $n - 2$ is produced. It remains to be shown that no word is produced by two different trees and that every word is generated from some tree. We shall achieve both ends by showing that the mapping has an inverse; i.e., for every word there is a unique tree which produces it.

Let $w = a_1a_2 \ldots a_{n-2}$ be a word over V. If T is a tree for which the algorithm produces w then the degree of vertex k, $d(k)$, in T, is equal to the number of times k appears in w, plus 1. This follows from the observation that when each, but the last, of the edges incident to k is deleted, k is written as a letter of w; the last edge may never be deleted, if k is one of the two vertices remaining in the tree, or if it is deleted, k is now the removed leaf, and the adjacent vertex, not k, is the written letter. Thus, if w is produced by the algorithm, for some tree, then the degrees of the vertices in the tree must be as stated.

For example, if $w = 4164$ then $d(1) = 2$, $d(2) = 1$, $d(3) = 1$, $d(4) = 3$, $d(5) = 1$ and $d(6) = 2$ in a tree which produced w.

Given this data, apply the following algorithm:

(1) $i \leftarrow 1$.
(2) Let j be the least vertex for which $d(j) = 1$. Construct an edge j—a_i, $d(j) \leftarrow 0$ and $d(a_i) \leftarrow d(a_i) - 1$.
(3) If $i = n - 2$, construct an edge between the two vertices whose degree is 1 and stop.
(4) Increment i and go to step 2.

It is easy to see that this algorithm picks the same vertex j as the original algorithm, and constructs a tree (the proof is by induction). Also, each step of the reconstruction is forced, therefore it is the only tree which yields w, and for every word this algorithm produces a tree.

In our example, for $i = 1, j = 2$ and since $a_1 = 4$ we connect 2—4, as shown in Figure 2.4. Now, $d(1) = 2$, $d(2) = 0$, $d(3) = 1$, $d(4) = 2$, $d(5) = 1$ and $d(6) = 2$. For $i = 2, j = 3$ and since $a_2 = 1$ we connect 3—1, as shown in Figure 2.5. Now $d(1) = 1$, $d(2) = 0$, $d(3) = 0$, $d(4) = 2$, $d(5) = 1$ and $d(6) = 2$. For $i = 3, j = 1$ and since $a_3 = 6$ we connect 1—6 as shown in Figure 2.6. Now, $d(1) = d(2) = d(3) = 0$, $d(4) = 2$ and $d(5) = d(6) = 1$. Finally, $i = 4, j = 5$ and since $a_4 = 4$ we connect 5—4, as shown in Figure 2.7. Now, $d(1) = d(2) = d(3) = d(5) = 0$ and $d(4) = d(6) = 1$. By step 3, we connect 4—6 and stop. The resulting graph is as in Figure 2.2.

Q.E.D.

A similar problem, stated and solved by Lempel and Welch [6], is that of finding the number of ways m labeled (distinct) edges can be joined by unlabeled endpoints to form a tree. Their proof is along the lines of Prüfer's proof of Cayley's theorem and is therefore constructive, in the sense that one can use the inverse transformation to generate all the trees after the words are generated. However, a much simpler proof was pointed out to me by A. Pnueli and is the subject of Problem 2.5.

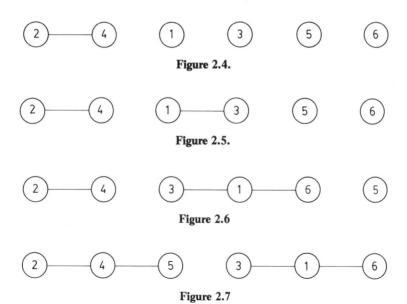

Figure 2.4.

Figure 2.5.

Figure 2.6

Figure 2.7

2.4 DIRECTED TREE DEFINITIONS

A digraph $G(V, E)$ is said to have a *root* r if $r \in V$ and every vertex $v \in V$ is *reachable* from r; i.e., there is a directed path which starts in r and ends in v.

A digraph (finite or infinite) is called a *directed tree* if it has a root and its underlying undirected graph is a tree.

Theorem 2.5: Assume G is a digraph. The following five conditions are equivalent:

(a) G is a directed tree.

(b) G has a root from which there is a unique directed path to every vertex.

(c) G has a root r for which $d_{in}(r) = 0$ and for every other vertex v, $d_{in}(v) = 1$.

(d) G has a root and the deletion of any edge (but no vertices) interrupts this condition.

(e) The underlying undirected graph of G is connected and G has one vertex r for which $d_{in}(r) = 0$, while for every other vertex v, $d_{in}(v) = 1$.

Proof: We prove that (a) $\Rightarrow$ (b) $\Rightarrow$ (c) $\Rightarrow$ (d) $\Rightarrow$ (e) $\Rightarrow$ (a).

(a) $\Rightarrow$ (b): We assume that G has a root, say r, and its underlying undirected graph G' is a tree. Thus, by Theorem 2.1, part (c), there is a unique simple path from r to every vertex in G'; also, G' is circuit-free. Thus, a directed path from r to a vertex v, in G, must be simple and unique.

(b) $\Rightarrow$ (c): Here we assume that G has a root, say r, and a unique directed path from it to every vertex v. First, let us show that $d_{in}(r) = 0$. Assume there is an edge $u \xrightarrow{e} r$. There is a directed path from r to u, and it can be continued, via e, back to r. Thus, in addition to the empty path from r to itself (containing no edges), there is one more, in contradiction of the assumption of the path uniqueness. Now, we have to show that if $v \neq r$ then $d_{in}(v) = 1$. Clearly, $d_{in}(v) > 0$ for it must be reachable from r. If $d_{in}(v) > 1$, then there are at least two edges, say $v_1 \xrightarrow{e_1} v$ and $v_2 \xrightarrow{e_2} v$. Since there is a directed path P_1 from r to v_1, and a directed path P_2 from r to v_2, by adding e_1 to P_1 and e_2 to P_2 we get two different paths from r to v. (This proof is valid even if $v_1 = v_2$.)

(c) $\Rightarrow$ (d): This proof is trivial, for the deletion on any edge $u \xrightarrow{e} v$ will make v unreachable from r.

(d) ⇒ (e): We assume that G has a root, say r, and the deletion of any edge interrupts this condition. First $d_{in}(r) = 0$, for any edge entering r could be deleted without interrupting the condition that r is a root. For every other vertex v, $d_{in}(v) > 0$, for it is reachable from r. If $d_{in}(v) > 1$, let $v_1 \xrightarrow{e_1} v$ and $v_2 \xrightarrow{e_2} v$ be two edges entering v. Let P be a simple directed path from r to v. It cannot use both e_1 and e_2. The one which is not used in P can be deleted without interrupting the fact that r is a root. Thus, $d_{in}(v) = 1$.

(e) ⇒ (a): We assume that the underlying undirected graph of G, G', is connected, $d_{in}(r) = 0$ and for $v \neq r$, $d_{in}(v) = 1$. First let us prove that r is a root. Let P' be a simple path connecting r and v in G'. This must correspond to a directed path P from r to v in G, for if any of the edges points in the wrong direction it would either imply that $d_{in}(r) > 0$ or that for some u, $d_{in}(u) > 1$. Finally, G' must be circuit-free, for a simple circuit in G' must correspond to a simple directed circuit in G (again using $d_{in}(r) = 0$ and $d_{in}(v) = 1$ for $v \neq r$), and at least one of its vertices, u, must have $d_{in}(u) > 1$, since the vertices of the circuit are reachable from r.

Q.E.D.

In case of finite digraphs one more useful definition of a directed tree is possible:

Theorem 2.6: A finite digraph G is a directed tree if and only if its underlying undirected graph, G', is circuit-free, one of its vertices, r, satisfies $d_{in}(r) = 0$, and for all other vertices v, $d_{in}(v) = 1$.

Proof: The "only if" part follows directly from the definition of a directed tree and Theorem 2.5, part (c).

To prove the "if" part we first observe that the number of edges is $n - 1$. Thus, by Theorem 2.2, (b) ⇒ (c), G' is connected. Thus, by Theorem 2.5, (e) ⇒ (a), G is a directed tree.

Q.E.D.

Let us say that a digraph is *arbitrated* (Berge [7] calls it quasi strongly connected) if for every two vertices v_1 and v_2 there is a vertex v_1 called an *arbiter* of v_1 and v_2, such that there are directed paths from v to v_1 and from v to v_2. There are infinite digraphs which are arbitrated but do not have a root. For example, see the digraph of Figure 2.8. However, for finite digraphs the following theorem holds:

Figure 2.8

Theorem 2.7: If a finite digraph is arbitrated then it has a root.

Proof: Let $G(V, E)$ be a finite arbitrated digraph, where $V = \{1, 2, \ldots, n\}$. Let us prove, by induction, that every set $\{1, 2, \ldots, m\}$, where $m \leq n$, has an arbiter; i.e., a vertex a_m such that every $1 \leq i \leq m$ is reachable from a_m. By definition, a_2 exists. Assume a_{m-1} exists. Let a_m be the arbiter of a_{m-1} and m. Since a_{m-1} is reachable from a_m and every $1 \leq i \leq m - 1$ is reachable from a_{m-1}, every $1 \leq i \leq m - 1$ is also reachable from a_m.

Q.E.D.

Thus, for finite digraphs, the condition that it has a root, as in Theorem 2.5 part *a, b, c* and *d*, can be replaced by it being arbitrated.

2.5   THE INFINITY LEMMA

The following is known as König's Infinity Lemma [8]:

Theorem 2.8: If G is an infinite digraph, with a root r and finite out-degrees for all its vertices, then G has an infinite directed path, starting in r.

Before we present the proof let us point out the necessity of the finiteness of the out-degrees of the vertices. For if we allow a single vertex to be of infinite out-degree, the conclusion does not follow. Consider the digraph of Figure 2.9. The root is connected to vertices v_1^1, v_1^2, v_1^3, $\ldots$, where v_1^k is the second vertex on a directed path of length k. It is clear that the tree is infinite, and yet it has no infinite path. Furthermore, the replacement of the condition of finite degrees by the condition that for every k the tree has a path of length k, does not work either, as the same example shows.

Proof: First let us restrict our attention to a directed tree T which is an infinite subgraph of G. T's root is r. All vertices of distance 1 away from r in G are also of distance 1 away from r in T. In general, if a vertex v is of distance l away from r in G it is also of distance l away from r in T; all the edges entering v in G are now dropped, except one which connects a vertex of distance $l - 1$ to v. It is sufficient to show that in T there is an infinite directed path from r. Clearly, since T is a subgraph of G, all its vertices are of finite outdegrees too.

In T, r has infinitely many descendants (vertices reachable from r). Since r is of finite out-degree, at least one of its sons (the vertices reachable via one edge), say r_1, must have infinitely many descendants. One of r_1's sons

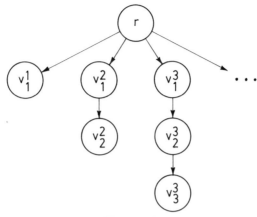

Figure 2.9

has infinitely many descendants, too, and so we continue to construct an infinite directed path $r, r_1, r_2, \ldots$.

Q.E.D.

In spite of the simplicity of the theorem, it is useful. For example, if we conduct a search on a directed tree of finite degrees (where a bound on the degree may not be known) for which it is known that it has no infinite directed paths, then the theorem ensures us that the tree is finite and our search will terminate.

An interesting application of Theorem 2.8 was made by Wang [9]. Consider the problem of tiling the plane with square tiles, all of the same size (Wang calls the tiles "dominoes"). There is a finite number of tile families. The sides of the tiles are labeled by letters of an alphabet, and all the tiles of one family have the same labels, thus are indistinguishable. Tiles may not be rotated or reflected, and the labels are specified for their north side, south side, and so on. There is an infinite supply of tiles of each family. The tiles may be put one next to another, the sides converging only if these two sides have the same labels. For example, if the tile families are as shown in Figure 2.10, then we can construct the "torus" shown in Figure 2.11. Now, by repeating this torus infinitely many times horizontally and vertically, we can tile the whole plane.

Wang proved that if it is possible to tile the upper right quadrant of the plane with a given finite set of tile families, then it is possible to tile the whole plane. The reader should realize that a southwest shift of the upper-right tiled quadrant cannot be used to cover the whole plane. In fact, if the number of tile families is not restricted to be finite, one can find sets

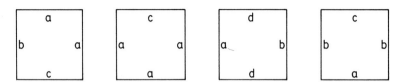

Figure 2.10

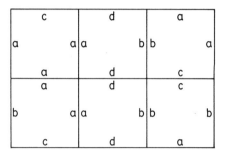

Figure 2.11

of families for which the upper-right quadrant is tileable, while the whole plane is not.

Consider the following directed tree T: The root r is connected to vertices, each representing one of the tile families, i.e., a square 1×1 tiled with the tile of that family. For every k, each one of the legitimate ways of tiling a $(2k + 1) \times (2k + 1)$ square is represented by a vertex in T; its father is the vertex which represents the tiling of a $(2k - 1) \times (2k - 1)$ square, identical to the center part of the square represented by the son.

Now, if the upper-right quadrant is tilable, then T has infinitely many vertices. Since the number of families is finite, the out-degree of each vertex is finite (although, may not be bounded). By Theorem 2.8, there is an infinite directed path in T. Such a path describes a way to tile the whole plane.

2.6 THE NUMBER OF SPANNING TREES

A subgraph H of a finite digraph G is called a *directed spanning tree* of G if H is a directed tree which includes all the vertices of G. If r is the root of H, then it is clearly a root of G. Also, if r is a root of G, then a spanning directed tree H of G exists with root r. This is simply observed by constructing H, edge by edge, starting from r and adding each time an edge of G from a vertex already reachable from r in H to one which is not reachable yet.

We shall now describe a method of Tutte [10] for computing the number of spanning directed trees of a given digraph with a given specified root. (For historical details, see reference 11.)

Let us define the **in-degree matrix D** of a digraph $G(V, E)$, where $V = \{1, 2, \ldots, n\}$, as follows:

$$D(i, j) = \begin{cases} d_{in}(i) & \text{if } i = j, \\ -k & \text{if } i \neq j, \text{ where } k \text{ is the number of edges in } G \text{ from } i \text{ to } j. \end{cases}$$

Lemma 2.1: A finite digraph $G(V, E)$, with no self-loops is a directed tree with root r if and only if its in-degree matrix D has the following two properties:

(1) $D(i, i) = \begin{cases} 0 & \text{if } i = r, \\ 1 & \text{if } i \neq r. \end{cases}$

(2) The minor, resulting from erasing the rth row and column from D and computing the determinant, is 1.

Proof: Assume that $G(V, E)$ is a directed tree with root r. By Theorem 2.5, part c, D satisfies property (1). Now, renumber the vertices in such a way that 1 is the root and if $i \to j$ then $i < j$. This can be achieved by numbering the vertices of unit distance from 1 as 2, 3, ... Next number the vertices of distance two, three, etc. The new in-degree matrix is derivable from the previous one by performing some permutation on the rows, and the same permutation on the columns. Since such a permutation does not change the determinant, the two minors are the same. The new in-degree matrix D' satisfies the following properties:

$$D'(1, 1) = 0,$$
$$D'(i, i) = 1 \quad \text{for} \quad i = 2, 3, \ldots, n,$$
$$D'(i, j) = 0 \quad \text{if} \quad i > j.$$

Thus, the minor, resulting from the erasure of the first row and the first column from D' and computing the determinant, is 1.

Now assume that D satisfies properties (1) and (2). By property (1) and Theorem 2.6, if G is not a directed tree then its underlying undirected graph contains a simple circuit. The vertex r cannot be one of the vertices of the circuit, for this would imply that either $d_{in}(r) > 0$ or for some other vertex v $d_{in}(v) > 1$, contrary to property (1). The circuit must be of the form:

$$i_1 \rightarrow i_2 \rightarrow \cdots \rightarrow i_l \rightarrow i_1,$$

where l is the length of the circuit, and no vertex appears on it twice. Also, there may be other edges out of i_1, i_2, $\ldots$, i_l, but none can enter. Thus, each of the columns of D, corresponding to one on this vertices, has exactly one $+1$ (on the main diagonal of D) and one -1 and all the other entries are 0. Also, each of the rows of this submatrix is either all zeros, or there is one $+1$ and one -1. The sum of these columns is therefore a zero column, and thus, the minor is 0. This contradicts property (2).

Q.E.D.

As a side result of our proof, we have the additional property that the minor of a graph whose in-degree matrix satisfies property (1) is 0 if the graph is not a directed tree with root r.

Theorem 2.9: The number of directed spanning trees with root r of a digraph with no self-loops is given by the minor of its in-degree matrix which results from the erasure of the rth row and column.

The proof of this theorem follows immediately from Lemma 2.1, the comment following it, and the linearity of the determinant function with respect to its columns. Let us demonstrate this by the following example. Consider the graph shown in Figure 2.12. Its in-degree matrix D is as follows:

$$D = \begin{bmatrix} 2 & -1 & -1 \\ -1 & 1 & -2 \\ -1 & 0 & 3 \end{bmatrix}$$

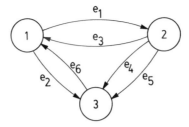

Figure 2.12

Assume that we want to compute the number of directed spanning trees with root 2. We erase the second row and column. The resulting determinant is

$$\begin{vmatrix} 2 & -1 \\ -1 & 3 \end{vmatrix} = 5$$

Now let us decompose this determinant into columns which represent one edge in every column. First, the 2×2 determinant can be written as

$$\begin{vmatrix} 2 & 0 & -1 \\ -1 & 1 & -2 \\ -1 & 0 & 3 \end{vmatrix}$$

We have returned the second row of D except its second entry, which must be made equal to 1 (in this case its value did not change). All other entries in the second column are changed into zero. Next, we decompose each column, except the second, into columns which consist of a single $+1$ and a single -1, as follows:

$$\begin{vmatrix} 2 & 0 & -1 \\ -1 & 1 & -2 \\ -1 & 0 & 3 \end{vmatrix} = \begin{vmatrix} 1 & 0 & -1 \\ -1 & 1 & -2 \\ 0 & 0 & 3 \end{vmatrix} + \begin{vmatrix} 1 & 0 & -1 \\ 0 & 1 & -2 \\ -1 & 0 & 3 \end{vmatrix}$$

$$= \begin{vmatrix} 1 & 0 & -1 \\ -1 & 1 & 0 \\ 0 & 0 & 1 \end{vmatrix} + \begin{vmatrix} 1 & 0 & 0 \\ -1 & 1 & -1 \\ 0 & 0 & 1 \end{vmatrix}$$

$$+ \begin{vmatrix} 1 & 0 & 0 \\ -1 & 1 & -1 \\ 0 & 0 & 1 \end{vmatrix} + \begin{vmatrix} 1 & 0 & -1 \\ 0 & 1 & 0 \\ -1 & 0 & 1 \end{vmatrix}$$

$$+ \begin{vmatrix} 1 & 0 & 0 \\ 0 & 1 & -1 \\ -1 & 0 & 1 \end{vmatrix} + \begin{vmatrix} 1 & 0 & 0 \\ 0 & 1 & -1 \\ -1 & 0 & 1 \end{vmatrix}$$

These six determinants correspond to the following selections of sets of edges, respectively: $\{e_3, e_2\}$, $\{e_3, e_4\}$, $\{e_3, e_5\}$, $\{e_6, e_2\}$, $\{e_6, e_4\}$, $\{e_6, e_5\}$. After erasing the second row and column, this corresponds to

$$\begin{vmatrix} 2 & -1 \\ -1 & 3 \end{vmatrix} = \begin{vmatrix} 1 & -1 \\ 0 & 1 \end{vmatrix} + \begin{vmatrix} 1 & 0 \\ 0 & 1 \end{vmatrix} + \begin{vmatrix} 1 & 0 \\ 0 & 1 \end{vmatrix}$$

$$+ \begin{vmatrix} 1 & -1 \\ -1 & 1 \end{vmatrix} + \begin{vmatrix} 1 & 0 \\ -1 & 1 \end{vmatrix} + \begin{vmatrix} 1 & 0 \\ -1 & 1 \end{vmatrix}$$

Each of these six determinants corresponds to a selection of $n - 1$ edges of the original graph. By Lemma 2.1, the resulting subgraph is a directed tree with root 2 if and only if the corresponding determinant is equal to one. Otherwise, it is zero. Thus, we get the number of directed trees with 2 as a root. Clearly, in our case, the only set which does not yield a directed tree is $\{e_6, e_2\}$ and indeed

$$\begin{vmatrix} 1 & -1 \\ -1 & 1 \end{vmatrix} = 0.$$

Q.E.D.

Let us now consider the question of the number of spanning trees of a given undirected graph $G(V, E)$. Consider the digraph $G'(V, E')$ defined as follows: For every edge $u \overset{e}{—} v$ in G define the two edges $u \overset{e'}{\rightarrow} v$ and $v \overset{e''}{\rightarrow} u$ in G'. Let r be a vertex. There is a one-one correspondence between the set of spanning trees of G and the set of directed spanning trees of G' with root r: Let T be a spanning tree of G. If the edge $u \overset{e}{—} v$ is in T and if u is closer than v to r in T then pick e' for T'; if v is closer, pick e''. Also, given T', it is easy to find the corresponding T by simply ignoring the directions; i.e., the existence of either e' or e'' in T' implies that e is in T. Thus, we can compute the number of spanning trees of G by writing the in-degree matrix of G', and computing the minor with respect to r. Clearly, the choice of r cannot make any difference. Now, the in-degree matrix of G' is given by

$$
D(i, j) = \begin{cases} d(i) & \text{in } G \text{ if } i = j, \\ -k & \text{where } k \text{ is the number of edges connecting } i \text{ and } j \\ & \text{in } G. \end{cases}
$$

This matrix is called the *degree matrix* of G. Hence, we have the following theorem:

Theorem 2.10: The number of spanning trees of an undirected graph with no self-loops is equal to any of the minors of its degree matrix which results from the erasure of a row and a corresponding column.

We can now use Theorem 2.10 to describe another proof of Cayley's Theorem (2.4). The number of spanning trees that can be constructed with vertices $1, 2, \ldots, n$ is equal to the number of spanning trees of the *complete* graph of n vertices; that is, the graph $G(V, E)$ with $V = \{1, 2, \ldots, n\}$ and for every $i \neq j$ there is one edge $i—j$. Its degree matrix is

$$
\begin{bmatrix} n-1 & -1 & \cdots & -1 \\ -1 & n-1 & & -1 \\ \vdots & & & \vdots \\ -1 & -1 & \cdots & n-1 \end{bmatrix}.
$$

After erasing one row and the corresponding column, the matrix looks the same, except that it is now $(n-1) \times (n-1)$. We can now add to any column (or row) a linear combination of the others, without changing its determinant. First subtract the first column from every other. We get:

$$
\begin{bmatrix}
n-1 & -n & -n & \cdots & -n \\
-1 & n & 0 & & 0 \\
-1 & 0 & n & & 0 \\
\vdots & & & & \vdots \\
-1 & 0 & 0 & & n
\end{bmatrix}
$$

Now add every one of the other rows to the first:

$$
\begin{bmatrix}
1 & 0 & 0 & \cdots & 0 \\
-1 & n & 0 & & 0 \\
-1 & 0 & n & & 0 \\
\vdots & \vdots & \vdots & & \vdots \\
-1 & 0 & 0 & & n
\end{bmatrix}
$$

Clearly, the determinant of this matrix is n^{n-2}.

2.7 OPTIMUM BRANCHINGS AND DIRECTED SPANNING TREES

A subgraph $B(V, E')$ of a finite digraph $G(V, E)$ is called a *branching* if it is circuit-free and $d_{in}(v) \leq 1$ for every $v \in V$. Clearly, if for only one vertex r, $d_{in}(r) = 0$ and for all the rest of the vertices, v, $d_{in}(v) = 1$ then, by Theorem 2.6, the branching is a directed tree with root r.

Let each edge e have a cost $c(e)$. Our problem is to find a branching $B(V, E')$ for which the sum of the edge costs, $\Sigma_{e \in E'} c(e)$, is maximum. This problem was solved independently by a number of authors [12, 13, 14]. We shall follow here Karp's paper [15]. First, we will show how to find a maximum branching. Then, we shall point out the simple modification for finding a minimum branching and a minimum spanning tree.

Let us call an edge $u \xrightarrow{e} v$, of G, *critical* if

(i) $c(e) > 0$ and
(ii) for all other edges $u' \xrightarrow{e'} v$, $c(e) \geq c(e')$.

Let H be a set of critical edges, where for each vertex one entering critical edge is chosen, if any exist. The graph (V, H) is called *critical*.

Lemma 2.2: If a critical graph (V, H) is circuit-free then it is a maximum branching.

Proof: Clearly (V, H) is a branching if it is circuit-free. If vertex v has no positive edges entering it in G and if B is a branching, then either B has no edge entering v, or if it has one, we can drop it without reducing B's total cost. Clearly, H contains no edge which enters v either. If vertex v has positive edges which enter it in G, then the one in H is of maximum cost, and therefore no branching can do better here either. Since H is at least as good as B in each vertex, (V, H) is a maximum branching.

Q.E.D.

If a critical graph contains circuits then it is not a branching. Let us study some of its properties.

Lemma 2.3: Each vertex in a critical graph is on at most one circuit.

Proof: If a vertex v is on two directed circuits then there must be a vertex u for which $d_{in}(u) \geq 2$; a contradiction. Such a vertex can be found by tracing backwards on one of the circuits which passes through v.

Q.E.D.

Let $B(V, E')$ be a branching and $u \xrightarrow{e} v$ an edge not in B. Then e is *eligible* relative to B if the set

$$E'' = E' \cup \{e\} - \{e' \mid e' \in E' \text{ and it enters } v\}$$

yields a branching (V, E'').

Lemma 2.4: Let $B(V, E')$ be a branching and $e \in E - E'$. $u \xrightarrow{e} v$ is eligible relative to B if and only if there is no directed path in B from v to u.

Proof: If there is a directed path from v to u in B, then when we add e a directed circuit is formed. The deletion of the edge entering v in B, if any, cannot open this circuit. Thus, e is not eligible.

If there is no directed path from v to u in B, then the addition of e cannot close a directed circuit. However, the resulting edge set may not be a branching if B already has an edge entering v. Now this edge is deleted, and the resulting edge set is a branching. Thus, e is eligible.

Q.E.D.

Lemma 2.5: Let B be a branching and C be a directed circuit such that no edge of $C - B$ is eligible relative to B. Then $|C - B| = 1$.

Proof: Since B is circuit-free, $|C - B| \geq 1$. Let $e_1, e_2, \ldots, e_k$ be the edges in $C - B$, in the order in which they appear on C. More specifically

$$u_1 \xrightarrow{e_1} v_1 \xrightarrow{p_1} u_2 \xrightarrow{e_2} v_2 \xrightarrow{p_2} \cdots \xrightarrow{p_{k-1}} u_k \xrightarrow{e_k} v_k \xrightarrow{p_k} u_1$$

is the circuit C, where the p_i's are directed paths common to C and B. Since e_1 is not eligible, by Lemma 2.4 there is a directed path in B, from v_1 to u_1. This path leaves p_1 somewhere and enters v_k, and continues via p_k to u_1; it cannot enter p_k after v_k, or B would have two edges entering the same vertex. Similarly, e_j is not eligible, and therefore there is a directed path in B, from v_j to u_j, which leaves p_j somewhere and enters p_{j-1} at v_{j-1}. We now have a directed circuit in B: It goes from v_1, via part of p_1, to a path leading to v_k, via part of p_k, to a path leading to v_{k-1}, etc., until it returns to v_1. The situation is described pictorially in Figure 2.13. Since B is circuit-free, $k \leq 1$.

Q.E.D.

Theorem 2.11: Let (V, H) be a critical graph. There exists a maximum branching $B(V, E')$ such that for every directed circuit C in (V, H), $|C - E'| = 1$.

Proof: Let $B(V, E')$ be a maximum branching which, among all maximum branchings, contains a maximum number of edges of (V, H). Consider $e \in H - E'$. If $u \xrightarrow{e} v$ is eligible then

$$E' \cup \{e\} - \{e' \mid e' \in E' \text{ and it enters } v\}$$

defines another maximum branching which contains one more edge of (V, H), a contradiction. Thus, no edge of $H - E'$ is eligible. By Lemma 2.5, for every directed circuit C in (V, H), $|C - E'| = 1$.

Q.E.D.

Thus, we can restrict our search of maximum branchings to those which share with (V, H) all the edges on circuits, except one edge per circuit.

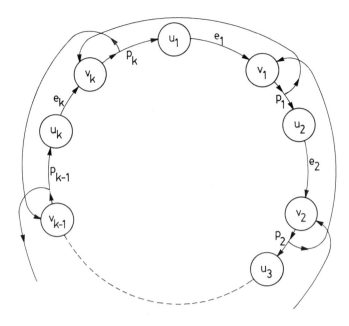

Figure 2.13

If B contains an edge $u \xrightarrow{e} v$ such that u is not on C but v is, then the edge entering v in C must be dropped. But if no such edge exists, then an edge e^0 on C, whose cost is minimum is dropped. This suggests the following method: We shrink each circuit C into a single vertex a and solve the maximum branching problem on the new digraph. We then expand this branching to be a branching of the original graph by dropping from each circuit one edge; if in the new digraph the branching contains no edge entering a then we drop from C the edge e^0; if the branching contains an edge entering a which corresponds to $u \xrightarrow{e} v$ in G then we drop from C the edge which enters v. It remains to define the new graph, including the cost of its edges and to prove that indeed the resulting branching is maximum in G.

Let $C_1, C_2, \ldots, C_k$ be the circuits of the critical graph (V, H). Let e_i^0 be an edge of minimum cost on C_i and let V_i be the set of vertices on C_i. Define $\hat{V} = V - \cup_{i=1}^k V_i$. For $u \xrightarrow{e} v$, where $u \notin V_i$ but $v \in V_i$, let $\bar{e}$ be the edge of C_i which enters v. We define the shrunk digraph $\bar{G}(\bar{V}, \bar{E})$ as follows:

$\bar{V} = \hat{V} \cup \{a_1, a_2, \ldots, a_k\}$, where the a_i's are new symbols.

$\bar{E} = \{u \xrightarrow{e} v \,|\, \text{either } u \text{ and } v \text{ are in } \hat{V} \text{ and } e \in E$
        or $u \in \hat{V}$, $v = a_i$ and $u \xrightarrow{e} w$ in G where w is on C_i,
        or $u = a_i$, $v \in \hat{V}$ and $w \xrightarrow{e} v$ in G, where w is on C_i,
        or $u = a_i$, $v = a_j$, $i \neq j$ and $w \xrightarrow{e} x$ in G where w is on C_i and
        x is on C_j.$\}$

$$\bar{c}(e) = \begin{cases} c(e) & \text{if } u \xrightarrow{e} v \text{ in } G \text{ and } v \in \hat{V}, \\ c(e) - c(\bar{e}) + c(e_i^0) & \text{if } u \xrightarrow{e} v \text{ in } G \text{ and } v \text{ is on } C_i. \end{cases}$$

The reason for this definition of $\bar{c}(e)$, in case v is on C_i, is that the least cost we must drop is $c(e_i^0)$. If we do pick e as part of the branching, we gain $c(e)$, but we lose $c(\bar{e}) - c(e_i^0)$, since now $\bar{e}$ must be dropped instead of e_i^0.

Let $\mathfrak{B}$ be the set of all branchings $B(V, E')$ of G for which $|C_i - E'| = 1$ for every circuit C_i of the critical graph (V, H) and if there is no edge $u \xrightarrow{e} v$ in B, with $u \notin V_i$ and $v \in V_i$ then $C_i - E' = \{e_i^0\}$. By Theorem 2.11, $\mathfrak{B}$ contains some maximum branchings. Also, let $\bar{\mathfrak{B}}$ be the set of all branchings of $\bar{G}$.

Theorem 2.12: There is a one-one correspondence between $\mathfrak{B}$ and $\bar{\mathfrak{B}}$. The branching $\bar{B}(\bar{V}, \bar{E}') \in \bar{\mathfrak{B}}$ which corresponds to $B(V, E') \in \mathfrak{B}$ is defined by $\bar{E}' = E' \cap \bar{E}$ and

$$c(B) = \bar{c}(\bar{B}) + \sum_{i=1}^{k} c(C_i) - \sum_{i=1}^{k} c(e_i^0) \tag{1}$$

where $c(B)$ and $\bar{c}(\bar{B})$ are the total costs of B and $\bar{B}$ respectively and $c(C_i)$ is the total cost of the edges in C_i.

Proof: First assume that $B(V, E') \in \mathfrak{B}$ and let us show that $\bar{B}(\bar{V}, \bar{E}')$ defined by $\bar{E}' = E \cap \bar{E}$ is a branching of $\bar{G}$. In B there can be at most one edge entering V_i from outside; this follows from the fact that all edges of C_i, but one, are in B and therefore only one vertex can have one edge enter it from $V - V_i$. Thus, in $\bar{B}$ there can be at most one edge entering a_i. If $v \in \hat{V}$, then in B and therefore in $\bar{B}$, there can be only one edge entering v. It remains to be shown that $\bar{B}$ is circuit-free. A directed circuit in $\bar{B}$ would indicate a directed circuit in B; whenever the circuit in $\bar{B}$ goes through a vertex a_i, in B we go around C_i and exit from the vertex u, where e is the next edge on the circuit in $\bar{B}$ and $u \overset{e}{\to} v$ in G. As we go around C_i, the missing edge is never needed, because in B we enter C_i at the vertex whose incoming edge in C_i is the missing one.

Next, assume that $\bar{B}(\bar{V}, \bar{E}')$ is a branching of $\bar{G}$ and let us show that there is exactly one $B(V, E') \in \mathfrak{B}$ such that $\bar{E}' = E' \cap \bar{E}$. For all edges $e \in \bar{E}'$ such that $u \overset{e}{\to} v$ and $v \in \hat{V}$, put e in E'. For all edges $e \in \bar{E}'$ such that $u \overset{e}{\to} a_i$, put e in E', and all the edges of C_i, except $\bar{e}$. If there is no edge $u \overset{e}{\to} a_i$ in $\bar{B}$ then put in E' all the edges of C_i except e_i^0. Clearly, in the resulting graph (V, E'), $d_{in}(v) \le 1$ for every $v \in V$, and if it contains a directed circuit, then there must be one in $\bar{B}$ too. Thus, it is a branching and $\bar{E}' = E' \cap \bar{E}$.

We have established the one-one correspondence and it remains to be shown that (1) holds. There are three cases in B to be considered:

(1) $u \overset{e}{\to} v$ and $v \in \hat{V}$. In this case e is both in B and $\bar{B}$ and $c(e) = \bar{c}(e)$. Thus, e contributes the same amount to both sides of (1).

(2) $u \overset{e}{\to} v$ where $u \notin V_i$ and $v \in V_i$. Again e is both in B and $\bar{B}$, but $c(e) = \bar{c}(e) + c(\bar{e}) - c(e_i^0)$. Clearly, in this case all edges of C_i, except $\bar{e}$ are in B. Thus, the total contribution of edges entering V_i to $c(B)$ is

$$\bar{c}(e) + c(C_i) - c(e_i^0).$$

(3) There is no edge $u \overset{e}{\to} a_i$ in $\bar{B}$. In this case there is no edge $u \overset{e}{\to} v$, $u \notin V_i$, $v \in V_i$ in B. However, all edges of C_i except e_i^0 are in B, contributing $c(C_i) - c(e_i^0)$.

We conclude that (1) holds.

Q.E.D.

The resulting algorithm is of time complexity $O(|E| \cdot |V|)$, since finding a critical graph is $O(|E|)$ and this may have to be repeated at most $|V|$ times. It is not hard to see that all the additional computations and bookkeeping does not increase the complexity beyond this bound. For more efficient algorithms see Tarjan [16].

The case of minimum branching is easily solved by a similar algorithm. All we have to do is change (i) to $c(e) < 0$ and in (ii) require $c(e) \le c(e')$, in the definition of a critical graph. If we want to find a directed spanning tree with root r, assuming that r is a root of G, we remove (i) altogether from the definition of a critical graph and remove all edges entering r in G.

2.8 DIRECTED TREES AND EULER CIRCUITS

Let $G(V, E)$ be a finite digraph which satisfies $d_{out}(v) = d_{in}(v)$ for every $v \in V$ and whose undirected underlying graph is connected. By Theorem 1.2 G has Euler paths, and all of them are circuits. Assume that

$$C: v_{i_1} \xrightarrow{e_1} v_{i_2} \xrightarrow{e_2} \ldots v_{im-1} \xrightarrow{em-1} v_{im} \xrightarrow{em} v_{i_1}$$

is one such circuit. Thus, if $i \neq j$, then $e_i \neq e_j$, and $m = |E|$; but vertices may repeat. Consider now a subgraph $H_c(V, E')$ defined in the following way: Let e_{j_1} be any one of the edges entering v_1. Let $|V| = n$. For every $p = 2, 3, \ldots, n$ let e_{j_p} be the first edge on C to enter v_p after the appearance of e_{j_1}. Now $E' = \{e_{j_2}, e_{j_3}, \ldots, e_{j_n}\}$.

For example, consider the graph of Figure 2.14.

The sequence of edges $e_1, e_2, \ldots, e_6$ designates a Euler circuit C. If we choose $e_{j_1} = e_6$ (the only choice in this case), then $e_{j_2} = e_1$, $e_{j_3} = e_2$ and $e_{j_4} = e_4$. The resulting subgraph $H_c(V, E')$ is shown in Figure 2.15, and is easily observed to be a directed spanning tree of G. The following lemma states this fact in general.

Lemma 2.6: Let C be a directed Euler circuit of a directed finite graph $G(V, E)$. The subgraph $H_c(V, E')$, constructed as described above, is a directed spanning tree of G, with root v_1.

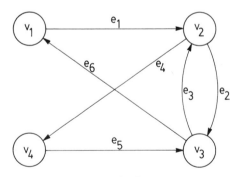

Figure 2.14

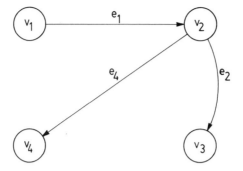

Figure 2.15

Proof: The definition of H implies that $d_{in}(v_1) = 0$ while $d_{in}(v_p) = 1$ for $p = 2, 3, \ldots, n$. By Theorem 2.6 it remains to be shown that the underlying undirected graph of H is circuit-free. It suffices to show that H has no directed circuits, but an edge $u \overset{e}{\to} v$ in H means that u is discovered before v, if we go along C starting from the reference edge e_{j_1}. Thus, no directed circuits exist in H.

Q.E.D.

Now assume that we are given a directed spanning tree H of a finite digraph G for which $d_{out}(v) = d_{in}(v)$ for every $v \in V$. Assume also that v_1 is the root of H and e_{j_1} is one of the edges entering v_1 in G. We shall describe now a method for using this information to trace an Euler circuit of G, without having to add detours, in contrast to the approach of section 1.3. We start from v_1 and leave it backwards on e_{j_1}. In general, upon arrival at v_p we leave it backwards via an edge entering v_p which was never used before, and we use e_{j_p} only when no other choice remains. If no unused entering edges exist, we stop.

By the construction, we never use an edge more than once. Also, the only vertex in which the process can stop is v_1 itself; this is guaranteed by the fact that $d_{out}(v) = d_{in}(v)$ for every vertex v, and the situation is like that of Section 1.3. When the process stops, all the edges emanating from v_1, and among those the edges which belong to H, have been traced back. Thus, all the vertices of distance 1 from v_1 in H have been visited; furthermore, since the edges of H have been used, all other incoming edges to them have been used, too. Thus, all vertices of distance 2 from v_1 in H have been visited; and so on. By induction, all edges of G have been traced.

In the construction of a Euler circuit from H and e_{j_1}, there are places of choice. If $d_{in}(v_p) > 1$, there are $(d_{in}(v_p) - 1)!$ different orders for picking the incoming edges (with e_{j_p} picked last). Also, it is clear that different orders will yield different Euler circuits. Thus, the number of distinct Euler circuits to be constructed from a given H and e_{j_1} is

$$\prod_{p=1}^{n} (d_{in}(v_p) - 1)!.$$

Furthermore, different choices of H (but with the same root v_1 and the same e_{j_1}) yield different Euler circuits; because a different e_{j_p} for some $2 \leq p \leq n$, will yield a different first entry to v_p after e_{j_1} in the resulting Euler circuit.

Finally, Lemma 2.6 guarantees that every Euler circuit will be generated for some H and some choice of ordering the backtracking edges, because the construction of Euler circuits from a directed tree is the reversal of the procedure of deriving a directed tree from a circuit.

We have thus proved the following theorem:

Theorem 2.13: The number of Euler circuits of a given finite digraph $G(V, E)$ for which $d_{out}(v) = d_{in}(v)$ for every $v \in V$ and whose undirected underlying graph is connected, is given by

$$\Delta \cdot \prod_{p=1}^{n} (d_{in}(v_p) - 1)!$$

where Δ is the number of directed spanning trees of G.

Clearly the result cannot depend on the choice of the root. This proves that if $d_{out}(v) = d_{in}(v)$ for every $v \in V$ then the number of spanning trees is the same for every choice of root.

For example, the in-degree matrix of the digraph of Figure 2.14 is

$$D = \begin{bmatrix} 1 & -1 & 0 & 0 \\ 0 & 2 & -1 & -1 \\ -1 & -1 & 2 & 0 \\ 0 & 0 & -1 & 1 \end{bmatrix}$$

By Theorem 2.9,

$$\Delta = \begin{bmatrix} 2 & -1 & -1 \\ -1 & 2 & 0 \\ 0 & -1 & 1 \end{bmatrix} = 2$$

which in this case is also the number of Euler circuits.

The reader is cautioned that if G contains self-loops, one should remove them before using Theorem 2.9, but of course take them into account in Theorem 2.13.

PROBLEMS

2.1 Let $T_1(V, E_1)$ and $T_2(V, E_2)$ be two spanning trees of $G(V, E)$. Prove that for every $\alpha \in E_1 \cap \bar{E}_2$ there exists a $\beta \in \bar{E}_1 \cap E_2$ such that each of the sets

$$(E_1 - \{\alpha\}) \cup \{\beta\}$$
$$(E_2 - \{\beta\}) \cup \{\alpha\}$$

defines a spanning tree.

2.2 Let $G(V, E)$ and $l(e)$ be as in the discussion of the minimum spanning tree problem. Give an algorithm for finding a maximum-length spanning tree and explain briefly why it works and its complexity.

2.3 Show that if $d(v_1), d(v_2), \ldots, d(v_n)$ are positive integers which satisfy the condition

$$\sum_{i=1}^{n} d(v_i) = 2n - 2$$

then there exists a tree with $v_1, v_2, \ldots, v_n$ as vertices and the d's specify the vertices' degree in the tree. How many different trees are there (if edges are unlabeled)?

2.4 Compute the number of trees that can be built on n given labeled vertices, with unlabeled edges, in such a way that one specified vertex is of degree k.

2.5 What is the number of trees that one can build with n labeled vertices and $m = n - 1$ labeled edges? Prove that the number of trees that can be built with m labeled edges (and no labels on the vertices) is $(m + 1)^{m-2}$.

2.6 Prove the following: If G is an infinite undirected connected graph, whose vertices are of finite degrees, then every vertex G is the start vertex of some simple infinite path.

2.7 Show that if rotation or flipping of tiles is allowed, the question of tiling the plane becomes trivial.

2.8 Describe a method for computing the number of in-going directed trees of a given digraph with a designated root. (Here a root is a vertex r such that from every vertex there is a directed path to r.) Explain why the method is valid. (Hint: Define an out-degree matrix.)

2.9 How many directed trees with root 000 are there in $G_{2,4}$?

Our purpose in Problems 2.10, 2.11 and 2.12 is to develop deBruijn's formula for the number of deBruijn sequences for the words of length n over an alphabet of σ letters.

2.10 Show that $\Delta_{\sigma,n}$, the number of spanning trees of $G_{\sigma,n}$ satisfies

$$\Delta_{\sigma,n} = \Delta_{\sigma,n-1} \cdot \sigma^{\sigma^{n-1} - \sigma^{n-2} - 1}.$$

Hint: Form $D_{\sigma,n}$, the D in-matrix of $G_{\sigma,n}$. Erase its first row and column. Subtract the last σ^{n-2} rows from the first $\sigma^{n-1} - \sigma^{n-2} - 1$ rows, to eliminate all the -1's. Add the first $\sigma^{n-1} - \sigma^{n-2} - 1$ columns to others to factor out $\sigma^{\sigma^{n-1} - \sigma^{n-2} - 1}$. In the remaining determinant add all rows to the first.

2.11 By Problem 2.10 show that $\Delta_{\sigma,n} = \sigma^{\sigma^{n-1} - n}$.

2.12 Show that the number of deBruijn sequences for a given σ and n is

$$\frac{(\sigma!)^{\sigma^{n-1}}}{\sigma^n}$$

(A similar development of deBruijn's formula can be found in Knuth [17].)

2.13 Let T be an undirected tree with n vertices. We want to invest $O(n)$ time labeling the graph properly, in such a way that once the labeling is done, for every two vertices, of distance l apart in the tree, one can find the (minimum) path connecting them in time $O(l)$. Describe briefly the preparatory algorithm, and the algorithm for finding the path.

2.14 Prove that the complement of a tree (contains the same vertices and an edge between two vertices iff no such edge exists in the tree) is either connected, or consists of one isolated vertex while all the others form a clique (a graph in which every pair of vertices is connected by an edge).

2.15 Let $G(V, E)$ be an undirected finite graph, where each edge e has a given length $l(e) > 0$. Let $\lambda(v)$ be the length of a shortest path from s to v.

(a) Explain why each vertex $v \neq s$ has an incident edge $u \xrightarrow{e} v$ such that $\lambda(v) = \lambda(u) + l(e)$.

(b) Show that if such an edge is chosen for each vertex $v \neq s$ (as in (a)) then the set of edges forms a spanning tree of G.

(c) Is this spanning tree always of minimum total weight? Justify your answer.

REFERENCES

1. Prim, R. C., "Shortest Connection Networks and Some Generalizations," *Bell System Tech. J.,* Vol. 36, 1957, pp. 1389-1401.

2. Cheriton, D., and Tarjan, R. E., "Finding Minimum Spanning Trees," *SIAM J. on Comp.,* Vol. 5, No. 4, Dec. 1976, pp. 724-742.

3. Cayley, A., "A Theorem on Trees," *Quart. J. Math.,* Vol. 23, 1889, pp. 376-378, *Collected Papers,* Vol. 13, Cambridge, 1897, pp. 26-28.

4. Prüfer, H., "Neuer Beweis eines Satzes über Permutationen," *Arch. Math. Phys.,* Vol. 27, 1918, pp. 742-744.

5. Moon, J. W., "Various Proofs of Cayley's Formula for Counting Trees," *A Seminar Graph Theory,* F. Harary (ed.), Holt, Rinehart and Winston, 1967, pp. 70-78.

6. Lempel, A., and Welch, L. R., "Enumeration of Arc-Labelled Trees," Dept. of E.E., Univ. of Southern Cal, 1969.

7. Berge, C., and Ghouila-Houri, A., *Programming, Games and Transportation Networks,* Wiley, 1965, Sec. 7.4.

8. König, D., *Theorie der endlichen und unendlichen Graphen,* Liepzig, 1936, reprinted by Chelsea, 1950.

9. Wang, H., "Proving Theorems by Pattern Recognition, II," *Bell System Tech. J.,* Vol. 40, 1961, pp. 1-41.

10. Tutte, W. T., "The Dissection of Equilateral Triangles into Equilateral Triangles," *Proc. Cambridge Phil. Soc.,* Vol. 44, 1948, pp. 463-482.

11. Harary, F., *Graph-Theory,* Addison Wesley, 1969, Chap. 16.

12. Chu, Y. J., and T. H. Liu, "On the Shortest Arborescence of a Directed Graph," *Sci. Sinica,* 14, 1965, pp. 1396-1400.

13. Edmonds, J., "Optimum Branchings," *J. of Res. of the Nat. Bureau of Standards,* 71B, 1967, pp. 233-240.

14. Bock, F., "An Algorithm to Construct a Minimum Directed Spanning Tree in a Directed Network," *Developments in Operations Research,* Gordon and Breach, 1971, pp. 29-44.

15. Karp, R. M., "A Simple Derivation of Edmonds' Algorithm for Optimum Branchings," *Networks,* 1, 1971, pp. 265-272.

16. Tarjan, R. E., "Finding Optimum Branchings," *Networks,* 7, 1977, pp. 25-35.

17. Knuth, D. E., "Oriented Subtrees of an Arc Digraph," *J. Comb. Th.,* Vol. 3, 1967, pp. 309-314.

Chapter 3

DEPTH-FIRST SEARCH

3.1 DFS OF UNDIRECTED GRAPHS

The Depths-First Search technique is a method of scanning a finite undirected graph. Since the publication of the papers of Hopcroft and Tarjan [1, 2] it is widely recognized as a powerful technique for solving various graph problems. However, the algorithm is not new; it was already known in the 19th century as a technique for threading mazes. For example, see Lucas' [3] report of Trémaux's work. Another algorithm, which was suggested later by Tarry [4], is just as good for threading mazes, and in fact DFS is a special case of it; but it is the additional structure of DFS which makes it so useful.

Assume we are given a finite connected graph $G(V, E)$. Starting in one of the vertices we want to "walk" along the edges, from vertex to vertex, visit all the vertices and halt. We seek an algorithm that will guarantee that we scan the whole graph, and recognize when we are done, without wondering too long in the "maze". We allow no preplanning, as by studying the road-map before we start our excursion; we must make our decisions, one at a time, since we discover the structure of the graph as we scan it. Clearly, we need to leave some "markers" as we go along, to recognize the fact that we have returned to a place visited before. Let us mark the *passages*, namely the connections of the edges to vertices. If the graph is presented by incidence lists then we can think of each of the two appearances of an edge in the incidence lists of its two endpoints as its two passages. It suffices to use two types of markers: F for the first passage used to enter the vertex, and E for any other passage when used to leave the vertex. No marker is ever erased or changed. As we shall prove later the following algorithm will terminate in the original starting vertex s, after scanning each edge once in each direction.

Trémaux's Algorithm:

(1) $v \leftarrow s$.
(2) If there are no unmarked passages in v, go to (4).

(3) Choose an unmarked passage, mark it E and traverse the edge to its other endpoint u. If u has any marked passages (i.e. it is not a new vertex) mark the passage, through which u has just been entered, by E, traverse the edge back to v, and go to Step (2). If u has no marked passages (i.e. it is a new vertex), mark the passage through which u has been entered by F, $v \leftarrow u$ and go to Step (2).

(4) If there is no passage marked F, halt. (We are back in s and the scanning of the graph is complete.)

(5) Use the passage marked F, traverse the edge to its other endpoint u, $v \leftarrow u$ and go to Step (2).

Let us demonstrate the algorithm on the graph shown in Figure 3.1. The initial value of v, the place "where we are" or the center of activity, is s. All passages are unlabelled. We choose one, mark it E and traverse the edge. Its other endpoint is a ($u = a$). None of its passages are marked, therefore we mark the passage through which a has been entered by F, the new center of activity is a ($v = a$), and we return to Step (2). Since a has two unmarked passages, assume we choose the one leading to b. The passage is marked E and the one at b is marked F since b is new, etc. The complete excursion is shown in Figure 3.1 by the dashed line.

Lemma 3.1: Trémaux's algorithm never allows an edge to be traversed twice in the same direction.

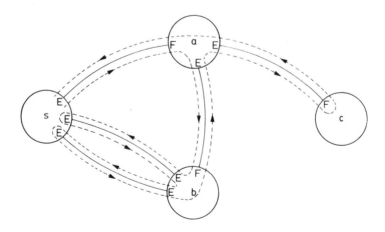

Figure 3.1

Proof: If a passage is used as an exit (entering an edge), then either it is being marked E in the process, and thus the edge is never traversed again in this direction, or the passage is already marked F. It remains to be shown that no passage marked F is ever reused for entering the edge.

Let $u \xrightarrow{e} v$ be the first edge to be traversed twice in the same direction, from u to v. The passage of e, at u, must be labeled F. Since s has no passages marked F, $u \neq s$. Vertex u has been left $d(u) + 1$ times; once through each of the passages marked E and twice through e. Thus, u must have been entered $d(u) + 1$ times and some edge $w \xrightarrow{e'} u$ has been used twice to enter u, before e is used for the second time. A contradiction.

Q.E.D.

An immediate corollary of Lemma 3.1 is that the process described by Trémaux's algorithm will always terminate. Clearly it can only terminate in s, since every other visited vertex has an F passage. Therefore, all we need to prove is that upon termination the whole graph has been scanned.

Lemma 3.2 Upon termination of Trémaux's algorithm each edge of the graph has been traversed once in each direction.

Proof: Let us state the proposition differently: For every vertex all its incident edges have been traversed in both directions.

First, consider the start vertex s. Since the algorithm has terminated, all its incident edges have been traversed from s outward. Thus, s has been left $d(s)$ times, and since we end up in s, it has also been entered $d(s)$ times. However, by Lemma 3.1 no edge is traversed more than once in the same direction. Therefore, all the edges incident to s have been traversed once in each direction.

Assume now that S is the set of vertices for which the statement, that each of their incident edges has been traversed once in each direction, holds. Assume $V \neq S$. By the connectivity of the graph there must be edges connecting vertices of S with $V - S$. All these edges have been traversed once in each direction. Let $v \xrightarrow{e} u$ be the first edge to be traversed from $v \in S$ to $u \in V - S$. Clearly, u's passage, corresponding to e, is marked F. Since this passage has been entered, all other passages must have been marked E. Thus, each of u's incident edges has been traversed outward. The search has not started in u and has not ended in u. Therefore, u has been entered $d(u)$ times, and each of its incident edges has been traversed inward. A contradiction, since u belongs in S.

Q.E.D.

The Hopcroft and Tarjan version of DFS is essentially the same as Trémaux's, except that they number the vertices from 1 to n ($=|V|$) in the order in which they are discovered. This is not necessary, as we have seen, for scanning the graph, but the numbering is useful in applying the algorithm for more advanced tasks. Let us denote the number of vertex v by $k(v)$. Also, instead of marking passages we shall now mark edges as "used" and instead of using the F mark to indicate the edge through which we leave the vertex for the last time, let us remember for each vertex v, other than s, the vertex $f(v)$ from which v has been discovered. $f(v)$ is called the *father* of v; this name will be justified later. DFS is now in the following form:

(1) Mark all the edges "unused". For every $v \in V$, $k(v) \leftarrow 0$. Also, let $i \leftarrow 0$ and $v \leftarrow s$.

(2) $i \leftarrow i + 1$, $k(v) \leftarrow i$.

(3) If v has no unused incident edges, go to Step (5).

(4) Choose an unused incident edge $v \overset{e}{\text{---}} u$. Mark e "used". If $k(u) \neq 0$, go to Step (3). Otherwise ($k(u) = 0$)), $f(u) \leftarrow v$, $v \leftarrow u$ and go to Step (2).

(5) If $k(v) = 1$, halt.

(6) $v \leftarrow f(v)$ and go to Step (3).

Since this algorithm is just a simple variation of the previous one, our proof that the whole (connected) graph will be scanned, each edge once in each direction, still applies. Here, in Step (4), if $k(u) \neq 0$ then u is not a new vertex and we "return" to v and continue from there. Also, moving our center of activity from v to $f(v)$ (Step (6)) corresponds to traversing the edge v—$f(v)$, in this direction. Thus, the whole algorithm is of time complexity $O(|E|)$, namely, linear in the size of the graph.

After applying the DFS to a finite and connected $G(V, E)$ let us consider the set of edges E' consisting of all the edges $f(v)$ —v through which new vertices have been discovered. Also direct each such edge from $f(v)$ to v.

Lemma 3.3: The digraph (V, E') defined above is a directed tree with root s.

Proof: Clearly $d_{in}(s) = 0$ and $d_{in}(v) = 1$ for every $v \neq s$. To prove that s is a root consider the sequence $v = v_0, v_1, v_2, \ldots$ where $v_{i+1} = f(v_i)$ for $i \geq 0$. Clearly, this defines a directed path leading into v in (V, E'). The path must be simple, since v_{i+1} was discovered before v_i. Thus, it can only terminate in s (which has no $f(s)$). Now by Theorem 2.5 (see part (c)), (V, E') is a directed tree with root s.

Q.E.D.

Clearly, if we ignore now the edge directions, (V, E') is a spanning tree of G. The following very useful lemma is due to Hopcroft and Tarjan [1, 2]:

Lemma 3.4: If an edge $a \overset{e}{-} b$ is not a part of (V, E') then a is either an ancestor or a descendant of b in the directed tree (V, E').

Proof: Without loss of generality, assume that $k(a) < k(b)$. In the DFS algorithm, the center of activity (v in the algorithm) moves only along the edges of the tree (V, E'). If b is not a descendant of a, and since a is discovered before b, the center of activity must first move from a to some ancestor of a before it moves up to b. However, we backtrack from a ($v \leftarrow f(a)$) only when all a's incident edges are used, which means that e is used and therefore b is already discovered—a contradiction.

Q.E.D.

Let us call all the edges of (V, E') *tree edges* and all the other edges *back edges*. The justification for this name is in Lemma 3.4; all the non-tree edges connect a vertex back to one of its ancestors.

Consider, as an example, the graph shown in Figure 3.2. Assume we start the DFS in c ($s = c$) and discover d, e, f, g, b, a in this order. The resulting vertex numbers, tree edges and back edges are shown in Figure 3.3, where the tree edges are shown by solid lines and are directed from low to high, and the back edges are shown by dashed lines and are directed from high to low. In both cases the direction of the edge indicates the direction in which the edge has been scanned first. For tree edges this is the defined direction, and for back edges we can prove it as follows: Assume $u \overset{e}{-} v$ is a back edge and u is an ancestor of v. The edge e could not have been scanned first from u, for if v has been undiscovered at that time then e would have been a tree edge, and if v has already been discovered (after u) then the center of activity could have been in u only if we have backtracked from v, and this means that e has already been scanned from v.

3.2 ALGORITHM FOR NONSEPARABLE COMPONENTS

A connected graph $G(V, E)$ is said to have a *separation vertex* v (sometimes also called an articulation point) if there exist vertices a and b, $a \neq v$ and $b \neq v$, such that all the paths connecting a and b pass through v. In this case we also say that v *separates* a from b. A graph which has a separation vertex is called *separable*, and one which has none is called *nonseparable*.

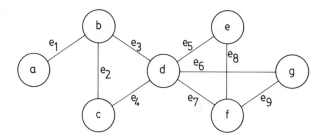

Figure 3.2

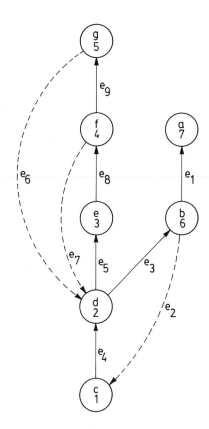

Figure 3.3

Let $V' \subseteq V$. The induced subgraph $G'(V', E')$ is called a *nonseparable component* if G' is nonseparable and if for every larger V'', $V' \subset V'' \subseteq V$, the induced subgraph $G''(V'', E'')$ is separable. For example, in the graph shown in Figure 3.2, the subsets $\{a, b\}$, $\{b, c, d\}$ and $\{d, e, f, g\}$ induce the nonseparable components of the graph.

If a graph $G(V, E)$ contains no separation vertex then clearly the whole G is a nonseparable component. However, if v is a separating vertex then $V - \{v\}$ can be partitioned into $V_1, V_2, \ldots, V_k$ such that $V_1 \cup V_2 \cup \ldots \cup V_k = V - \{v\}$ and if $i \neq j$ then $V_i \cap V_j = \emptyset$; two vertices a and b are in the same V_i if and only if there is a path connecting them which does not include v. Thus, no nonseparable component can contain vertices from more than one V_i. We can next consider each of the subgraphs induced by $V_i \cup \{v\}$ and continue to partition it into smaller parts if it is separable. Eventually, we end up with nonseparable parts. This shows that no two nonseparable components can share more than one vertex because each such vertex is a separating vertex. Also, every simple circuit of length greater than one must lie entirely in one nonseparable component.

Now, let us discuss how DFS can help to detect separating vertices.

Let the *lowpoint* of v, $L(v)$, be the least number, $k(u)$ of a vertex u which can be reached from v via a, possible empty, directed path consisting of tree edges followed by at most one back edge. Clearly $L(v) \leq k(v)$, for we can use the empty path from v to itself. Also, if a non-empty path is used then its last edge is a back edge, for a directed path of tree edges leads to vertices higher than v. For example, in the graph of Figure 3.2 with the DFS as shown in Figure 3.3 the lowpoints are as follows: $L(a) = 7$, $L(b) = L(c) = L(d) = 1$ and $L(e) = L(f) = L(g) = 2$.

Lemma 3.5: Let G be a graph whose vertices have been numbered by DFS. If $u \rightarrow v$ is a tree edge, $k(u) > 1$ and $L(v) \geq k(u)$ then u is a separating vertex of G.

Proof: Let S be the set of vertices on the path from the root r $(k(r) = 1)$ to u, including r but not including u, and let T be the set of vertices on the subtree rooted at v, including v (that is, all the descendants of v, including v itself). By Lemma 3.4 there cannot be any edge connecting a vertex of T with any vertex of $V - (S \cup \{u\} \cup T)$. Also, if there is any edge connecting a vertex $t \in T$ with a vertex $s \in S$ then the edge $t \rightarrow s$ is a back edge and clearly $k(s) < k(u)$. Now, $L(v) \leq k(s)$, since one can take the tree edges from v to t followed by $t \rightarrow s$. Thus, $L(v) < k(u)$, contradicting the hypothesis. Thus, u is separating the S vertices from the T vertices and is therefore a separating vertex.

Q.E.D.

Lemma 3.6: Let $G(V, E)$ be a graph whose vertices have been numbered by DFS. If u is a separating vertex and $k(u) > 1$ then there exists a tree edge $u \rightarrow v$ such that $L(v) \geq k(u)$.

Proof. Since u is a separating vertex, there is a partition of $V - \{u\}$ into V_1, $V_2, \ldots, V_m$ such that $m \geq 2$ and if $i \neq j$ then all paths from a vertex of V_i to a vertex of V_j pass through u. We assume that the search does not start in u. Let us assume it starts in r and $r \in V_1$. The center of activity of the DFS must pass through u. Let $u \rightarrow v$ be the first tree edge for which $v \notin V_1$. Assume $v \in V_2$. Since there are no edges connecting vertices of V_2 with $V - (V_2 \cup \{u\})$, $L(v) \geq k(u)$.

Q.E.D.

Lemma 3.7: Let $G(V, E)$ be a graph whose vertices have been numbered by DFS, starting with r $(k(r) = 1)$. The vertex r is a separating vertex if and only if there are at least two tree edges out of r.

Proof: Assume that r is a separating vertex. Let $V_1, V_2, \ldots, V_m$ be a partition of $V - \{r\}$ such that $m \geq 2$ and if $i \neq j$ then all paths from a vertex of V_i to a vertex of V_j pass through r. Therefore, no path in the tree which starts with $r \rightarrow v$, $v \in V_i$ can lead to a vertex of V_j where $j \neq i$. Thus, there are at least two tree edges out of r.

Now, assume $r \rightarrow v_1$ and $r \rightarrow v_2$ are two tree edges out of r. Let T be the set of vertices in the subtree rooted at v_1. By Lemma 3.4, there are no edges connecting vertices of T with vertices of $V - (T \cup \{r\})$. Thus, r separates T from the rest of the graph, which is not empty since it includes at least the vertex v_2.

Q.E.D.

Let $C_1, C_2, \ldots, C_m$ be the nonseparable components of the connected graph $G(V, E)$, and let $s_1, s_2, \ldots, s_p$ be its separating vertices. Let us define the *superstructure* of $G(V, E)$, $\tilde{G}(\tilde{V}, \tilde{E})$ as follows:

$$\tilde{V} = \{s_1, s_2, \ldots, s_p\} \cup \{C_1, C_2, \ldots, C_m\},$$

$$\tilde{E} = \{s_i - C_j \mid s_i \text{ is a vertex of } C_j \text{ in } G\}.$$

By the observations we have made in the beginning of the section, $\tilde{G}(\tilde{V}, \tilde{E})$ is a tree. By Corollary 2.1, if $m > 1$ then there must be at least two leaf components, each containing only one separating vertex, since for every separating vertex s_i $d(s_i) \geq 2$ in $\tilde{G}$. By Lemma 3.2, the whole graph will be explored by the DFS.

Now, assume the search starts in a vertex r which is not a separating vertex. Even if it is in one of the leaf-components, eventually we will enter another leaf-component C, say via its separating vertex u and the edge $u \overset{e}{\to} v$. By Lemma 3.6, $L(v) \geq k(u)$, and if $L(v)$ is known when we backtrack from v to u, then by using Lemma 3.5, we can detect that u is a separating vertex. Also, as far as the component C is concerned, from the time C is entered until it is entirely explored, we can think of the algorithm as running on C alone with u as the starting vertex. Thus, by Lemma 3.7, there will be only one tree edge from u into C, and all the other vertices of C are descendants of v and are therefore explored after v is discovered and before we backtrack on e. This suggests the use of a stack (pushdown store) for producing the vertices of the component. We store the vertices in the stack in the order that they are discovered. If on backtracking e we discover that u is a separating vertex, we read off all the vertices from the top of the stack down to and including v. All these vertices, plus u (which is not removed at this point from the stack even if it is the next on top) constitute the component. This, in effect removes the leaf C from the tree $\tilde{G}(\tilde{V}, \tilde{E})$, and if its adjacent vertex s (a separating vertex) has now $d(s) = 1$, then we can assume that it is removed too. The new superstructure is again a tree, and the same process will repeat itself to detect and trim one leaf at a time until only one component is left when the DFS terminates.

If the search starts in a separating vertex r, then all but the components which contain r are detected and produced as before, and the ones that do contain r are detected by Lemma 3.7: Each time we backtrack into r, on $r \to v$, if r has additional unexplored incident edges then we conclude that the vertices on the stack above and including v, plus r, constitute a component.

The remaining problem is that of computing $L(v)$ in time; i.e. its value should be known by the time we backtrack from v.

If v is a leaf of the DFS tree then $L(v)$ is the least element in the following set: $\{k(u) \mid u = v$ or $v \to u$ is a back edge$\}$. Let us assign $L(v) \leftarrow k(v)$ immediately when v is discovered, and as each back edge $v \to u$ is explored, let us assign

$$L(v) \to \text{Min}\{L(v), k(u)\}.$$

Clearly, by the time we backtrack from v, all the back edges have been explored, and $L(v)$ has the right value.

If v is not a leaf of the DFS tree, then $L(v)$ is the least element in the following set:

$$\{k(u) \mid u = v \text{ or } v \to u \text{ is a back edge}\} \cup$$

$$\{L(u) \mid v \to u \text{ is a tree edge}\}.$$

When we backtrack from v, we have already backtracked from all its sons earlier, and therefore already know their lowpoint. Thus, all we need to add is that when we backtrack from u to v, we assign

$$L(v) \leftarrow \text{Min}\{L(v), L(u)\}.$$

Let us assume that $|V| > 1$ and s is the vertex in which we start the search. The algorithm is now as follows:

(1) Mark all the edges "unused". Empty the stack S. For every $v \in V$ let $k(v) \leftarrow 0$. Let $i \leftarrow 0$ and $v \leftarrow s$.

(2) $i \leftarrow i + 1$, $k(v) \leftarrow i$, $L(v) \leftarrow i$ and put v on S.

(3) If v has no unused incident edges go to Step (5).

(4) Choose an unused incident edge $v \overset{e}{-} u$. Mark e "used". If $k(u) \neq 0$, let $L(v) \leftarrow \text{Min}\{L(v), k(u)\}$ and go to Step (3). Otherwise ($k(u) = 0$) let $f(u) \leftarrow v$, $v \leftarrow u$ and go to Step (2).

(5) If $k(f(v)) = 1$, go to Step (9).

(6) ($f(v) \neq s$). If $L(v) < k(f(v))$, then $L(f(v)) \leftarrow \text{Min}\{L(f(v)), L(v)\}$ and go to Step (8).

(7) ($L(v) \geq k(f(v))$) $f(v)$ is a separating vertex. All the vertices on S down to and including v are now removed from S; this set, with $f(v)$, forms a nonseparable component.

(8) $v \leftarrow f(v)$ and go to Step (3).

(9) All vertices on S down to and including v are now removed from S; they form with s a nonseparable component.

(10) If s has no unused incident edges then halt.

(11) Vertex s is a separating vertex. Let $v \leftarrow s$ and go to Step (4).

Although this algorithm is more complicated then the scanning algorithm, its time complexity is still $O(|E|)$. This follows easily from the fact that each edge is still scanned exactly once in each direction and that the number of operations per edge is bounded by a constant.

3.3 DFS ON DIGRAPHS

Let us consider now a DFS algorithm on digraphs. Again, each vertex v gets a number $k(v)$. The first scan of an edge $u \overset{e}{\to} v$ is always in the direction of the edge; if after e is scanned v is visited for the first time then $f(v) = u$. The algorithm is as follows:

(1) Mark all the edges "unused". For every $v \in V$ let $k(v) \leftarrow 0$ and $f(v)$ be "undefined". Also, let $i \leftarrow 0$ and $v \leftarrow s$. (s is the vertex we choose to start the search from.)
(2) $i \leftarrow i + 1$, $k(v) \leftarrow i$.
(3) If there are no unused incident edges from v then go to Step (5).
(4) Choose an unused edge $v \overset{e}{\to} u$. Mark e "used". If $k(u) \neq 0$, go to Step (3). Otherwise ($k(u) = 0$), $f(u) \leftarrow v$, $v \leftarrow u$ and go to Step (2).
(5) If $f(v)$ is defined then $v \leftarrow f(v)$ and go to Step (3).
(6) ($f(v)$ is undefined). If there is a vertex u for which $k(u) = 0$ then let $v \leftarrow u$ and go to Step (2).
(7) (All the vertices have been scanned) Halt.

The structure which results from the DFS of a digraph is not as simple as it is in the case of undirected graphs; instead of two types of edges (tree edges and back edges) there are four:

(i) *Tree edges*: An edge $x \overset{e}{\to} y$ is a tree edge if it was used to "discover" y; i.e. when it was scanned (Step (4)), $k(y) = 0$. Upon scanning e, the center of activity, v, shifted from x to y. Later, the center of activity returned to x (Step (5)) from y.
(ii) *Forward edges*: An edge $x \overset{e}{\to} y$ is a forward edge if when it was scanned for the first time, $k(y) \geq k(x)$. Since e was "unused", we had not backtracked from x yet. Thus, all the vertices with a higher label, including y, are descendants of x; i.e. are reachable from x via tree edges. Therefore, the forward edges do not add any newly reachable vertices and thus are usually ignored.
(iii) *Back edges*: An edge $x \overset{e}{\to} y$ is a back edge if when it was scanned for the first time, $k(y) < k(x)$ and y is an ancestor of x. (We shall need additional structure in order to be able to tell upon scanning e whether y is an ancestor of x or not.)
(iv) *Cross edges*: An edge $x \overset{e}{\to} y$ is a cross edge if when it was scanned for the first time, $k(y) < k(x)$ but y is not an ancestor of x.

Assume DFS was performed on a digraph $G(V, E)$, and let the set of tree edges be E'. The digraph (V, E') is a branching (Section 2.7), or as sometimes called, a forest, since it is a union of disjoint directed trees. The only remnant parallel of the structure of DFS for undirected graphs (as in Lemma 3.4) is in the fact that if $x \xrightarrow{e} y$ and $k(y) \geq k(x)$ then y is a descendant of x.

3.4 ALGORITHM FOR STRONGLY-CONNECTED COMPONENTS

Let $G(V, E)$ be a finite digraph. Let us define a relation $\sim$ on V. If $x, y \in V$ we say that $x \sim y$ if there is a directed path from x to y and also there is a directed path from y to x in G. Clearly $\sim$ is an equivalence relation, and it partitions V into equivalence classes. These equivalence classes are called the *strongly-connected components* of G.

Let $C_1, C_2, \ldots, C_m$ be the strongly-connected components of a digraph $G(V, E)$. Let us define the *superstructure* of G, $\tilde{G}(\tilde{V}, \tilde{E})$ as follows:

$$\tilde{V} = \{C_1, C_2, \ldots, C_m\}$$
$$\tilde{E} = \{C_i \xrightarrow{e} C_j \mid i \neq j, \, x \xrightarrow{e} y \text{ in } G, \, x \in C_i \text{ and } y \in C_j\}.$$

The digraph $\tilde{G}$ must be free of directed circuits; for if it has a directed circuit, all the strongly-connected components on it should have been one component. Thus, there must be at least one sink, C_k, i.e., $d_{\text{out}}(C_k) = 0$, in G. Let r be the first vertex of C_k visited in the DFS of $\tilde{G}$; r may have been reached via a tree edge $q \xrightarrow{e} r$, it may be s or it may have been picked by Step (6), in the last two cases it is a root of one of the trees of the forest. Now, all the vertices of C_k are reachable from r. Thus, no retracting from r is attempted until all the vertices of C_k are discovered; they all get numbers greater than $k(r)$, and since there are no edges out of C_k in G, no vertex is visited outside C_k form the time that r is discovered until we retract from it. Thus, if we store on a stack the vertices in the order of their discovery, then upon retraction from r, all the vertices on the stack, down to and including r are the elements of C_k. The only problem is how do we tell when we retract from a vertex that it has been the first one in a sink-component?

For this purpose let us again define the *lowpoint* of v, $L(v)$, to be the least number, $k(u)$, of a vertex u which can be reached from v via a, possibly empty, directed path consisting of tree edges followed by at most one back edge or a cross edge, provided u belongs to the same strongly-connected component. It seems like a circular situation; in order to compute the lowpoint we need to identify the components, and in order to find the components we need the lowpoint. However, Tarjan [2] found a way out. In the following algorithm, whose validity we shall discuss later, we use a stack S, on which we store the names of the vertices in the order in which they are discovered. Also, in an array, we record for each vertex whether it is on S or not, so that the question of whether it is on S can be answered is constant time. The algorithm is as follows:

(1) Mark all the edges "unused". For every $v \in V$ let $k(v) \leftarrow 0$ and $f(v)$ be "undefined". Empty S. Let $i \leftarrow 0$ and $v \leftarrow s$.

(2) $i \leftarrow i + 1$, $k(v) \leftarrow i$, $L(v) \leftarrow i$ and put v on S.

(3) If there are no unused incident edges from v then go to Step (7).

(4) Choose an unused edge $v \overset{e}{\to} u$. Mark e "used". If $k(u) = 0$ then $f(u) \leftarrow v$, $v \leftarrow u$ and go to Step (2).

(5) If $k(u) > k(v)$ (e is a forward edge) go to Step (3). Otherwise ($k(u) < k(v)$), if u is not on S (u and v do not belong to the same component) go to Step (3).

(6) ($k(u) < k(v)$ and both vertices are in the same component) Let $L(v) \leftarrow$ Min$\{L(v), k(u)\}$ and go to Step (3).

(7) If $L(v) = k(v)$ then delete all the vertices form S down to and including v; these vertices form a component.

(8) If $f(v)$ is defined then
$L(f(v)) \leftarrow$ Min$\{L(f(v)), L(v)\}$, $v \leftarrow f(v)$ and go to Step (3).

(9) ($f(v)$ is undefined) If there is a vertex u for which $k(u) = 0$ then let $v \leftarrow u$ and go to Step (2).

(10) (All vertices have been scanned) Halt.

Lemma 3.8: Let r be the first vertex for which, in Step (7), $L(r) = k(r)$. When this occurs, all the vertices on S, down to and including r, form a strongly-connected (sink) component of G.

Proof: All the vertices in S, on top of r, have been discovered after r, and since no backtracking from r has been tempted yet, these vertices are the descendants of r; i.e. are reachable from r via tree-edges.

Next, we want to show that if v is a descendant of r then r is reachable from v. We have already backtracked from v, but since r is the first vertex for which equality occurs in Step (7), $L(v) < k(v)$. Thus, a vertex u, $k(u) = L(v)$ is reachable from v. Also, $k(u) \geq k(r)$, since by Step (8) $L(r) \leq L(v)$, ($k(u) = L(v) \geq L(r) = k(r)$), and therefore u must be a descendant of r. If $u \neq r$ then we can repeat the argument again to find a lower numbered descendant of r reachable from u, and therefore from v. This argument can be repeated until r is found to be reachable from v.

So far we have shown that all the vertices on S, on top of r and including r, belong to the some component. It remains to be shown that no additional vertices belong to the same component.

When the equality $L(r) = k(r)$ is discovered, all the vertices with numbers higher than $k(r)$ are on top of r in S. Thus, if there are any additional vertices in the component, at least some of them must have been discovered already, and all of those are numberes lower than $k(r)$. There must be at least one (back or cross) edge $x \rightarrow y$ such that $k(x) \geq k(r) > k(y) > 0$ and x, r and y are in the same component. Since no removals from S have taken place yet, $L(x) \leq k(y)$ and therefore $L(r) \leq L(x) \leq k(y) < k(r)$, a contradiction.

Q.E.D.

If r is the first vertex for which in Step (7), $L(r) = k(r)$, then by Lemma 3.8, and its proof, a component C has been discovered and all its elements are descendants of r. Up to now at most one edge from a vertex outside C into a vertex in C may have been used, namely $f(r) \rightarrow r$. Thus, so far, no vertex of C has been used to change the lowpoint value of a vertex outside C. At this point all vertices of C are removed from S and therefore none of the edges entering C can change a lowpoint value anymore. Effectively this is equivalent to the removal of C and all its incident edges from G. Thus, when equality in Step (7) will occur again, Lemma 3.8 is effective again. This proves the validity of the algorithm.

PROBLEMS

3.1 Tarry's algorithm [3] is like Trémaux's, with the following change. Replace Step (3) by:

> (3) Choose an unmarked passage, mark it E and traverse the edge to its other endpoint u. If u has no marked passages (i.e. it is a new vertex), mark the passage through which u has been entered by F. Let $v \leftarrow u$ and go to Step (2).

Prove that Tarry's algorithm terminates after all the edges of G have been traversed, once in each direction. (Observe that Lemmas 3.1 and 3.2 remain valid for Tarry's algorithm, with virtually the same proofs.)

3.2 Consider the set of edges which upon the termination of Tarry's algorithm (see Problem 3.1) have one endpoint marked E and the other F; also assume these edges are now directed from E to F.

  (a) Prove that this set of edges is a directed spanning tree of G with root s. (See Lemma 3.3.).
  (b) Does a statement like that of Lemma 3.4 hold in this case? Prove or disprove.

3.3 Fraenkel [5, 6] showed that the number of edge traversals can sometimes be reduced if the use of a two-way counter is allowed. The algorithm is a variant of Tarry's algorithm (see Problem 3.1). Each time a new vertex is entered the counter is incremented; when it is realized that all incident edges of a vertex have been traversed at least in one direction, the counter is decremented. If the counter reaches the start value, the search is stopped. One can return to s via the F marked passages.

  Write an algorithm or a flow chart which realizes this idea. (Hint: an additional mark which temporarily marks the passages used to reenter a vertex is used.) Prove that the algorithm works. Show that for some graphs the algorithm will traverse each edge exactly once, for others the savings depends on the choice of passages, and yet there are graphs for which the algorithm cannot save any traversals.

3.4 Assume G is drawn in the plane in such a way that no two edges intersect. Show how Trémaux's algorithm can be modified in such a way that the whole scanning path never crosses itself.

3.5 In an undirected graph G a set of vertices C is called a *clique* if every two vertices of C are connected by an edge. Prove that in the spanning (directed) tree resulting from a DFS, all the vertices of a clique appear on one directed path. Do they necessarily appear consecutively on the path? Justify your answer.

3.6 Prove that if C is a directed circuit of a digraph to which a DFS algorithm was applied then the vertex v, for which $k(v)$ is minimum among the vertices of C, is a root of a subtree in the resulting forest, and all the vertices of C are in this subtree.

3.7 An edge e of a connected undirected graph G is called a *bridge* if its deletion destructs G's connectivity. Describe a variation of the DFS algorithm which, instead of detecting separating vertices, detects bridges.

3.8 (This problem was suggested by Silvio Micali) Let G be a connected graph.

  (a) Prove that a vertex $u \neq s$ is a separating vertex of G if and only if, upon the termination of DFS on G, there is a tree edge $u \rightarrow v$ for which there is no back edge $x \rightarrow y$ such that x is a descendant of v and y is a proper ancestor of u.

(b) Describe an algorithm, of complexity $O(|E|)$, which detects separating vertices and produces the nonseparable components without numbering of vertices, and therefore without the use of *lowpoint*. (Hint: For every back edge $x \rightarrow y$, mark all the tree edges on the path from y to x; proceed from x to y until an already marked edge is encountered or the tree edge outgoing from y is reached. This latter edge is not marked. The back edges are considered by rescanning G in the same order, but the marking for $x \rightarrow y$ is done when y is scanned for the first time. When all this is over, $v \neq s$ is a separating vertex if and only if there is an unmarked tree edge out of v.)

3.9 (This problem was suggested by Alessandro Tescari.) Show that the algorithm for nonseparable components can be simplified by adding a new edge $r - s$, where r is a new vertex, and starting the search at r. Do we still need Lemma 3.7?

REFERENCES

[1] Hopcroft, J., and Tarjan, R., "Algorithm 447: Efficient Algorithms for Graph Manipulation", Comm. ACM, Vol. 16, 1973, pp. 372-378.

[2] Tarjan, R., "Depth-First Search and Linear Graph Algorithms", SIAM J. Comput., Vol. 1, 1972, pp. 146-160.

[3] Lucas, E., *Récreations Mathématiques,* Paris, 1882.

[4] Tarry, G., "Le Problème des Labyrinthes". Nouvelles Ann. de Math., Vol. 14, 1895, page 187.

[5] Fraenkel, A. S., "Economic Traversal of Labyrinths", Math. Mag., Vol. 43, 1970, pp. 125-130.

[6] Fraenkel, A. S., "Economic Traversal of Labyrinths (Correction)," Math. Mag., Vol. 44, No. 1, January 1971.

Chapter 4

4. ORDERED TREES

4.1 UNIQUELY DECIPHERABLE CODES

Let $\Sigma = \{0, 1, \ldots, \sigma - 1\}$. We call Σ an *alphabet* and its elements are called *letters*; the number of letters in Σ is σ. (Except for this numerical use of σ, the "numerical" value of the letters is ignored; they are just "meaningless" characters. We use the numerals just because they are convenient characters.) A finite sequence $a_1 a_2 \cdots a_l$, where a_i is a letter, is called a *word* whose *length* is l. We denote the length of a word w by $l(w)$. A set of (non-empty and distinct) words is called a *code*. For example, the code $\{102, 21, 00\}$ consists of three code-words: one code-word of length 3 and two code-words of length 2; the alphabet is $\{0, 1, 2\}$ and consists of three letters. Such an alphabet is called *ternary*.

Let $c_1, c_2, \ldots, c_k$ be code-words. The *message* $c_1 c_2 \cdots c_k$ is the word resulting from the concatenation of the code-word c_1 with c_2, etc. For example, if $c_1 = 00$, $c_2 = 21$ and $c_3 = 00$, then $c_1 c_2 c_3 = 002100$.

A code C over Σ (that is, the code-words of C consist of letters in Σ) is said to be *uniquely decipherable* (UD) if every message constructed from code-words of C can be broken down into code-words of C in only one way. For example, the code $\{01, 0, 10\}$ is not UD because the message 010 can be parsed in two ways: 0, 10 and 01, 0.

Our first goal is to describe a test for deciding whether a given code C is UD. This test is an improvement of a test of Sardinas and Patterson [1] and can be found in Gallager's book [2].

If s, p and w are words and $ps = w$ then p is called a *prefix* of w and s is called a *suffix* of w. We say that a word w is non-empty if $l(w) > 0$.

A non-empty word t is called a *tail* if there exist two messages $c_1 c_2 \cdots c_m$ and $c_1' c_2' \cdots c_n'$ with the following properties:

(1) c_i, $1 \le i \le m$, and c_j', $1 \le j \le n$ are code-words and $c_1 \ne c_1'$;
(2) t is a suffix of c_n';
(3) $c_1 c_2 \cdots c_m t = c_1' c_2' \cdots c_n'$.

Lemma 4.1: A code C is UD if and only if no tail is a code-word.

Proof: If a code-word c is a tail then by definition there exist two messages $c_1 c_2 \cdots c_m$ and $c_1' c_2' \cdots c_n'$ which satisfy $c_1 c_2 \cdots c_m c = c_1' c_2' \cdots c_n'$, while $c_1 \neq c_1'$. Thus, there are two different ways to parse this message, and C is not UD.

If C is not UD then there exist messages which can be parsed in more than one way. Let μ be such an ambiguous message whose length is minimum: $\mu = c_1 c_2 \cdots c_k = c_1' c_2' \cdots c_n'$; i.e. all the c_i-s and c_j-s are code-words and $c_1 \neq c_1'$. Now, without loss of generality we can assume that c_k is a suffix of c_n' (or change sides). Thus, c_k is a tail.

Q.E.D.

The algorithm generates all the tails. If a code-word is a tail, the algorithm terminates with a negative answer.

Algorithm for UD:

(1) For every two code-words, c_i and c_j ($i \neq j$), do the following:

    (1.1) If $c_i = c_j$, halt; C is not UD.
    (1.2) If for some word s, either $c_i s = c_j$ or $c_i = c_j s$, put s in the set of tails.

(2) For every tail t and every code-word c do the following:

    (2.1) If $t = c$, halt; C is not UD.
    (2.2) If for some word s either $ts = c$ or $cs = t$, put s in the set of tails.

(3) Halt; C is UD.

Clearly, in Step (1), the words declared to be tails are indeed tails. In Step (2), since t is already known to be a tail, there exist code-words $c_1, c_2, \ldots, c_m$ and $c_1', c_2', \ldots, c_n'$ such that $c_1 c_2 \cdots c_m t = c_1' c_2' \cdots c_n'$. Now, if $ts = c$ then $c_1 c_2 \cdots c_m c = c_1' c_2' \cdots c_n' s$, and therefore s is a tail; and if $cs = t$ then $c_1 c_2 \cdots c_m c s = c_1' c_2' \cdots c_n'$ and s is a tail.

Next, if the algorithm halts in (3), we want to show that all the tails have been produced. Once this is established, it is easy to see that the conclusion that C is UD follows; Each tail has been checked, in Step (2.1), whether it is equal to a code-word, and no such equality has been found; by Lemma 4.1, the code C is UD.

For every t let $m(t) = c_1 c_2 \cdots c_m$ be a shortest message such that $c_1 c_2 \cdots c_m t = c_1' c_2' \cdots c_n'$, and t is a suffix of c_n'. We prove by induction on the length of $m(t)$ that t is produced. If $m(t) = 1$ then t is produced by (1.2), since $m = n = 1$.

Now assume that all tails p for which $m(p) < m(t)$ have been produced. Since t is a suffix of c_n', we have $pt = c_n'$. Therefore, $c_1 c_2 \cdots c_m = c_1' c_2' \cdots c_{n-1}' p$.

If $p = c_m$ then $c_m t = c_n'$ and t is produced in Step (1).

If p is a suffix of c_m then, by definition, p is a tail. Also, $m(p)$ is shorter then $m(t)$. By the inductive hypothesis p has been produced. In Step (2.2), when applied to the tail p and code-word c_n', by $pt = c_n'$, the tail t is produced.

If c_m is a suffix of p, then $c_m t$ is a suffix of c_n', and therefore, $c_m t$ is a tail. $m(c_m t) = c_1 c_2 \cdots c_{m-1}$, and is shorter than $m(t)$. By the inductive hypothesis $c_m t$ has been produced. In Step (2.2), when applied to the tail $c_m t$ and code-word c_m, the tail t is produced.

This proves that the algorithm halts with the right answer.

Let the code consists of n words and l be the maximum length of a code-word. Step (1) takes at most $O(n^2 \cdot l)$ elementary operations. The number of tails is at most $O(n \cdot l)$. Thus, Step (2) takes at most $O(n^2 l^2)$ elementary operations. Therefore, the whole algorithm is of time complexity $O(n^2 l^2)$. Other algorithms of the same complexity can be found in References 3 and 4; these tests are extendible to test for additional properties [5, 6, 7].

Theorem 4.1: Let $C = \{c_1, c_2, \ldots, c_n\}$ be a UD code over an alphabet of σ letters. If $l_i = l(c_i)$, $i = 1, 2, \ldots, n$, then

$$\sum_{i=1}^{n} \sigma^{-l_i} \leq 1. \tag{4.1}$$

The left hand side of (4.1) is called the *characteristic sum* of C; clearly, it characterizes the vector $(l_1, l_2, \ldots, l_n)$, rather than C. The inequality (4.1) is called the *characteristic sum condition*. The theorem was first proved by McMillan [8]. The following proof is due to Karush [9].

Proof: Let e be a positive integer

$$\left(\sum_{i=1}^{n} \sigma^{-l_i} \right)^e = \sum_{i_1=1}^{n} \sum_{i_2=1}^{n} \cdots \sum_{i_e=1}^{n} \sigma^{-(l_{i_1} + l_{i_2} + \cdots + l_{i_e})}.$$

There is a unique term, on the right hand side, for each of the n^e messages of e code-words. Let us denote by $N(e, j)$ the number of messages of e code-words whose length is j. It follows that

$$\sum_{i_1=1}^{n} \sum_{i_2=1}^{n} \cdots \sum_{i_e=1}^{n} \sigma^{-(l_{i_1} + l_{i_2} + \cdots + l_{i_e})} = \sum_{j=e}^{e\hat{l}} N(e, j) \cdot \sigma^{-j}$$

where $\hat{l}$ is the maximum length of a code-word. Since C is UD, no two messages can be equal. Thus, $N(e, j) \leq \sigma^j$. We now have,

$$\sum_{j=e}^{e \cdot l} N(e, j) \cdot \sigma^{-j} \leq \sum_{j=e}^{e \cdot l} \sigma^j \cdot \sigma^{-j} \leq e \cdot \hat{l}.$$

We conclude that for all $e \geq 1$

$$\left(\sum_{i=1}^{n} \sigma^{-l_i} \right)^e \leq e \cdot \hat{l}.$$

This implies (4.1).

Q.E.D.

A code C is said to be *prefix* if no code-word is a prefix of another. For example, the code $\{00, 10, 11, 100, 110\}$ is not prefix since 10 is a prefix of 100; the code $\{00, 10, 11, 010, 011\}$ is prefix. A prefix code has no tails, and is therefore UD. In fact it is very easy to parse messages: As we read the message from left to right, as soon as we read a code-word we know that it is the first code-word of the message, since it cannot be the beginning of another code-word. Therefore, in most applications, prefix codes are used. The following theorem, due to Kraft [10], in a sense, shows us that we do not need non-prefix codes.

Theorem 4.2: If the vector of integers, $(l_1, l_2, \ldots, l_n)$, satisfies

$$\sum_{i=1}^{n} \sigma^{-l_i} \leq 1 \tag{4.2}$$

then there exists a prefix code $C = \{c_1, c_2, \ldots, c_n\}$, over the alphabet of σ letters, such that $l_i = l(c_i)$.

Proof: Let $\lambda_1 < \lambda_2 < \cdots < \lambda_m$ be integers such that each l_i is equal to one of the λ_j-s and each λ_j is equal to at least one of the l_i-s. Let k_j be the number of l_j-s which are equal to λ_j. We have to show that there exists a prefix code C such that the number of code-words of length λ_j is k_j.

Clearly, (4.2) implies that

$$\sum_{j=1}^{m} k_j \sigma^{-\lambda_j} \leq 1 \tag{4.3}$$

We prove by induction on r that for every $1 \leq r \leq m$ there exists a prefix code C_r such that, for every $1 \leq j \leq r$, the number of its code-words of length λ_j is k_j.

First assume that $r = 1$. Inequality (4.3) implies that $k_1 \sigma^{-\lambda_1} \leq 1$, or $k_1 \leq \sigma^{\lambda_1}$. Since there are σ^{λ_1} distinct words of length λ_1, we can assign any k_1 of them to constitute C_1.

Now, assume C_r exists. If $r < m$ then (4.3) implies that

$$\sum_{j=1}^{r+1} k_j \sigma^{-\lambda_j} \leq 1.$$

Multiplying both sides by $\sigma^{\lambda_{r+1}}$ yields

$$\sum_{j=1}^{r+1} k_j \sigma^{\lambda_{r+1} - \lambda_j} \leq \sigma^{\lambda_{r+1}},$$

which is equivalent to

$$k_{r+1} \leq \sigma^{\lambda_{r+1}} - \sum_{j=1}^{r} k_j \sigma^{\lambda_{r+1} - \lambda_j}. \tag{4.4}$$

Out of the $\sigma^{\lambda_{r+1}}$ distinct words of length λ_{r+1}, $k_j \cdot \sigma^{\lambda_{r+1} - \lambda_j}$, $1 \leq j \leq r$, have prefixed of length λ_j as code-words of C_r. Thus, (4.4) implies that enough are left to assign k_{r+1} words of length λ_{r+1}, so that none has a prefix in C_r. The enlarged set of code-words is C_{r+1}.

Q.E.D.

This proof suggests an algorithm for the construction of a code with a given vector of code-word length. We shall return to the question of prefix code construction, but first we want to introduce positional trees.

4.2 POSITIONAL TREES AND HUFFMAN'S OPTIMIZATION PROBLEM

A positional σ-tree (or when σ is known, a positional tree) is a directed tree with the following property: Each edge out of a vertex v is associated with one of the letters of the alphabet $\Sigma = \{0, 1, \ldots, \sigma - 1\}$; different edges, out of v, are associated with different letters. It follows that the number of edges out of a vertex is at most σ, but may be less; in fact, a leaf has none.

We associate with each vertex v the word consisting of the sequence of letters associated with the edges on the path from the root r to v. For example, consider the binary tree (positional 2-tree) of Figure 4.1, where in each vertex the associated word is written. (λ denotes the empty word.)

Clearly, the set of words associated with the leaves of a positional tree is a prefix code. Also, every prefix code can be described by a positional tree in this way.

The *level* of a vertex v of a tree is the length of the directed path from the root to v; it is equal to the length of the word associated with v.

Our next goal is to describe a construction of an optimum code, in a sense to be discussed shortly. It is described here as a communication problem, as it was viewed by Huffman [11], who solved it. In the next section we shall describe one more application of this optimization technique.

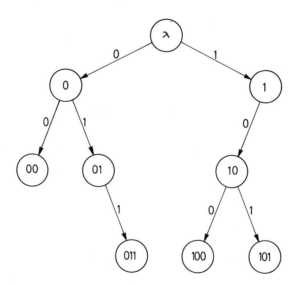

Figure 4.1

Assume words over a source alphabet of n letters have to be transmitted over a channel which can transfer one letter of the alphabet $\Sigma = \{0, 1, \ldots, \sigma - 1\}$ at a time, and $\sigma < n$. We want to construct a code over Σ with n code-words, and associate with each source letter a code-word. A word over the source alphabet is translated into a message over the code, by concatenating the code-words which correspond to the source letters, in the same order as they appear in the source word. This message can now be transmitted through the channel. Clearly, the code must be UD.

Assume further, that the source letters have given probabilities p_1, $p_2, \ldots, p_n$ of appearance, and the choice of the next letter in the source word is independent of its previous letters. If the vector of code-word lengths is $(l_1, l_2, \ldots, l_n)$ then the average code-word length, $\bar{l}$, is given by

$$\bar{l} = \sum_{i=1}^{n} p_i l_i. \tag{4.5}$$

We want to find a code for which $\bar{l}$ is minimum, in order to minimize the expected length of the message.

Since the code must be UD, by Theorem 4.1, the vector of code-word lengths must satisfy the characteristic sum condition. This implies, by Theorem 4.2, that a prefix code with the same vector of code-word lengths exists. Therefore, in seeking an optimum code, for which $\bar{l}$ is minimum, we may restrict our search to prefix codes. In fact, all we have to do is find a vector of code-word lengths for which $\bar{l}$ is minimum, among the vectors which satisfy the characteristic sum condition.

First, let us assume that $p_1 \geq p_2 \geq \cdots \geq p_n$. This is easily achieved by sorting the probabilities. We shall first demonstrate Huffman's construction for the binary case ($\sigma = 2$). Assume the probabilities are 0.6, 0.2, 0.05, 0.05, 0.03, 0.03, 0.03, 0.01. We write this list as our top row (see Fig. 4.2). We add the last (and therefore least) two numbers, and insert the sum in a proper place to maintain the non-increasing order. We repeat this operation until we get a vector with only two probabilities. Now, we assign each of them a word-length 1 and start working our way back up by assigning each of the probabilities of the previous step, its length in the present step, if it is not one of the last two, and each of the two last probabilities of the previous step is assigned a length larger by one than the length assigned to their sum in the present step.

Once the vector of code-word lengths is found, a prefix code can be assigned to it by the technique of the proof of Theorem 4.2. (An efficient implementation is discussed in Problem 4.6) Alternatively the back up pro-

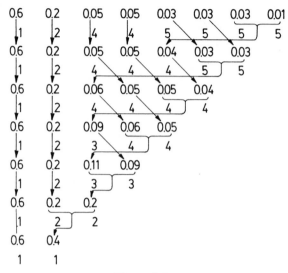

Figure 4.2

cedure can produce a prefix code directly. Instead of assigning the last two probabilities with lengths, we assign the two words of length one: 0 and 1. As we back up from a present step, in which each probability is already assigned a word, to the previous step, the rule is as follows: All, but the last two probabilities of the previous step are assigned the same words as in the present step. The last two probabilities are assigned $c0$ and $c1$, where c is the word assigned to their sum in the present step.

In the general case, when $\sigma \geq 2$, we add in each step the last d probabilities of the present vector of probabilities; if n is the number of probabilities of this vector then d is given by:

$$1 < d \leq \sigma \quad \text{and} \quad n \equiv d \bmod (\sigma - 1). \tag{4.6}$$

After the first step, the length of the vector, n', satisfies $n' \equiv 1 \bmod (\sigma - 1)$, and will be equal to one, mod $(\sigma - 1)$, from there on. The reason for this rule is that we should end up with exactly σ probabilities, each to be assigned length 1. Now, $\sigma \equiv 1 \bmod (\sigma - 1)$, and since in each ordinary step the number of probabilities is reduced by $\sigma - 1$, we want $n \equiv 1 \bmod (\sigma - 1)$. In case this condition is not satisfied by the given n, we correct it in the first step as is done by our rule. Our next goal is to prove that this indeed leads to an optimum assignment of a vector of code-word lengths.

Lemma 4.2: If $C = \{c_1, c_2, \ldots, c_n\}$ is an optimum prefix code for the probabilities $p_1, p_2, \ldots, p_n$ then $p_i > p_j$ implies that $l(c_i) \leq l(c_j)$.

Proof: Assume $l(c_i) > l(c_j)$. Make the following switch: Assign c_i to probability p_j, and c_j to p_i; all other assignments remain unchanged. Let $\tilde{l}$ denote the average code-word length of the new assignment, while $\bar{l}$ denotes the previous one. By (4.5) we have

$$
\begin{aligned}
\bar{l} - \tilde{l} &= [p_i \cdot l(c_i) + p_j \cdot l(c_j)] - [p_i \cdot l(c_j) + \\
&\quad p_j \cdot l(c_i)] \\
&= (p_i - p_j)(l(c_i) - l(c_j)) > 0,
\end{aligned}
$$

contradicting the assumption that $\bar{l}$ is minimum.

Q.E.D.

Lemma 4.3: There exists an optimum prefix code for the probabilities $p_1 \geq p_2 \geq \cdots \geq p_n$ such that the positional tree which represents it has the following properties:

(1) All the internal vertices of the tree, except possibly one internal vertex v, have exactly σ sons.
(2) Vertex v has $1 < \rho \leq \sigma$ sons, where $n \equiv \rho \mod (\sigma - 1)$.
(3) Vertex v, of (1) is on the lowest level which contains internal vertices, and its sons are assigned to $p_{n-\rho+1}, p_{n-\rho+2}, \ldots, p_n$.

Proof: Let T be a positional tree which represents an optimum prefix code. If there exists an internal vertex u, which is not on the lowest level of T containing internal vertices and it has less than σ sons, then we can perform the following change in T: Remove one of the leaves of T from its lowest level and assign to the probability a new son of u. The resulting tree, and therefore, its corresponding prefix code has a smaller average code-word length. A contradiction. Thus, we conclude that no such internal vertex u exists.

If there are internal vertices, on the lowest level of internal vertices, which have less than σ sons, choose one of them, say v. Now eliminate sons from v and attach their probabilities to new sons of the others, so that their number of sons is σ. Clearly, such a change does not change the average length and the tree remains optimum. If before filling in all the missing sons, v has no more sons, we can use v as a leaf and assign to it one of the probabilities from the lowest level, thus creating a new tree which is better than T. A contradiction. Thus, we never run out of sons of v to be transferred to other lacking internal vertices on the same level. Also, when this process ends, v is the only lacking internal vertex (proving (1)) and its number of remaining sons must be greater than one, or its son can be removed and its probability attached to v. This proves that the number of sons of v, ρ, satisfies $1 < \rho \leq \sigma$.

If v's ρ sons are removed, the new tree has $n' = n - \rho + 1$ leaves and is *full* (i.e., every internal vertex has exactly σ sons). In such a tree, the number of leaves, n', satisfies $n' \equiv 1$, mod $(\sigma - 1)$. This is easily proved by induction on the number of internal vertices. Thus, $n - \rho + 1 \equiv 1 \mod (\sigma - 1)$, and therefore $n \equiv \rho \mod (\sigma - 1)$, proving (2).

We have already shown that v is on the lowest level of T which contains internal vertices and number of its sons is ρ. By Lemma 4.2, we know that the least ρ probabilities are assigned to leaves of the lowest level of T. If they are not sons of v, we can exchange sons of v with sons of other internal vertices on this level, to bring all the least probabilities to v, without changing the average length.

$$\text{Q.E.D.}$$

For a given alphabet size σ and probabilities $p_1 \geq p_2 \geq \ldots \geq p_n$, let θ_σ $(p_1, p_2, \ldots, p_n)$ be the set of all σ-ary positional trees with n leaves, assigned with the probabilities $p_1, p_2, \ldots, p_n$ in such a way that p_{n-d+1}, $p_{n-d+2}, \ldots, p_n$ (see (4.6)) are assigned, in this order, to the first d sons of a vertex v, which has no other sons. By Lemma 4.3, $\theta_\sigma (p_1, p_2, \ldots, p_n)$ contains at least one optimum tree. Thus, we may restrict our search for an optimum tree to $\theta_\sigma (p_1, p_2, \ldots, p_n)$.

Lemma 4.4: There is a one to one correspondence between $\theta_\sigma(p_1, p_2, \ldots, p_n)$ and the set of σ-ary positional trees with $n - d + 1$ leaves assigned with $p_1, p_2, \ldots, p_{n-d}, p'$, where $p' = \sum_{i=n-d+1}^{n} p_i$. The average word-length $\bar{l}$, of the prefix code represented by a tree T of $\theta_\sigma(p_1, p_2, \ldots, p_n)$ and the average code word-length $\bar{l}'$, of the prefix code represented by the tree T', which corresponds to T, satisfy

$$\bar{l} = \bar{l}' + p'. \tag{4.7}$$

Proof: The tree T' which corresponds to T is achieved as follows: Let v be the father of the leaves assigned $p_{n-d+1}, p_{n-d+2}, \ldots, p_n$. Remove all the sons of v and assign p' to it.

It is easy to see that two different trees T_1 and T_2 in $\theta_\sigma (p_1, p_2, \ldots, p_n)$ will yield two different trees T_1' and T_2', and that every σ-ary tree T' with

$n - d + 1$ leaves assigned $p_1, p_2, \ldots, p_{n-d}, p'$, is the image of some T; establishing the correspondence.

Let l_i denote the level of the leaf assigned p_i in T. Clearly $l_{n-d+1} = l_{n-d+2} = \ldots = l_n$. Thus,

$$\bar{l} = \sum_{i=1}^{n-d} p_i \cdot l_i + l_n \cdot \sum_{i=n-d+1}^{n} p_i = \sum_{i=1}^{n-d} p_i l_i + l_n \cdot p'$$

$$= \sum_{i=1}^{n-d} p_i \cdot l_i + (l_n - 1) \cdot p' + p' = \bar{l}' + p'.$$

Q.E.D.

Lemma 4.4 suggests a recursive approach to find an optimum T. For $\bar{l}$ to be minimum, $\bar{l}'$ must be minimum. Thus, let us first find an optimum T' and then find T by attaching d sons to the vertex of T' assigned p'; these d sons are assigned $p_{n-d+1}, p_{n-d+2}, \ldots, p_n$. This is exactly what is done in Huffman's procedure, thus proving its validity.

It is easy to implement Huffman's algorithm in time complexity $O(n^2)$. First we sort the probabilities, and after each addition, the resulting probability is inserted in a proper place. Each such insertion takes at most $O(n)$ steps, and the number of insertions is $\lceil (n - \sigma)/(\sigma - 1) \rceil$. Thus, the whole forward process is of time complexity $O(n^2)$. The back up process is $O(n)$ if pointers are left in the forward process to indicate the probabilities of which it is composed.

However, the time complexity can be reduced to $O(n \log n)$. One way of doing it is the following: First sort the probabilities. This can be done in $O(n \log n)$ steps [14]. The sorted probabilities are put on a queue S_1 in a nonincreasing order from left to right. A second queue, S_2, initially empty, is used too. In the general step, we repeatedly take the least probability of the two (or one, if one of the queues is empty) appearing at the right hand side ends of the two queues, and add up d of them. The result, p', is inserted at the left hand side end of S_2. The process ends when after adding d probabilities both queues are empty. This adding process and the back up are $O(n)$. Thus, the whole algorithm is $O(n \log n)$.

The construction of an optimum prefix code, when the cost of the letters are not equal is discussed in Reference 12; the case of alphabetic prefix codes, where the words must maintain lexicographically the order of the given probabilities, is discussed in Reference 13. These references give additional references to previous work.

4.3 APPLICATION OF THE HUFFMAN TREE TO SORT-BY-MERGE TECHNIQUES

Assume we have n items, and there is an order defined between them. For ease of presentation, let us assume that the items are the integers $1, 2, \ldots, n$ and the order is "less than". Assume that we want to organize the numbers in nondecreasing order, where initially they are put in L lists, $A_1, A_2, \ldots,$ A_L. Each A_i is assumed to be ordered already. Our method of building larger lists from smaller ones is as follows. Let $B_1, B_2, \ldots, B_m$ be any m existing lists. We read the first, and therefore least, number in each of the lists, take the least number among them away from its list and put it as the first number of the merged list. The list from which we took the first number is now shorter by one. We repeat this operation on the same m lists until they merge into one. Clearly, some of the lists become empty before others, but since this depends on the structure of lists, we only know that the general step of finding the least number among m numbers (or less) and its transfer to a new list is repeated $b_1 + b_2 + \cdot + b_m$ times, where b_i is the number of numbers in B_i.

The number m is dictated by our equipment or decided upon in some other way. However, we shall assume that its value is fixed and predetermined. In fact, in most cases $m = 2$.

The whole procedure can be described by a positional tree. Consider the example shown in Fig. 4.3, where $m = 2$. First we merge the list $\langle 3 \rangle$ with $\langle 1, 4 \rangle$. Next we merge $\langle 2, 5 \rangle$ with $\langle 1, 3, 4 \rangle$. The original lists, $A_1, A_2, \ldots,$ A_L, correspond to the leaves of the tree. The number of transfers can be

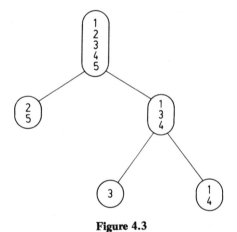

Figure 4.3

computed as follows: Let a_i be the number of numbers in A_i, and l_i be the level of the list A_i in the tree. The number of elementary merge operations is then

$$\sum_{i=1}^{L} a_i \cdot l_i. \tag{4.8}$$

Burge [15] observed that the attempt to find a positional m-ary which minimizes (4.8) is similar to that of the minimum average word-length problem solved by Huffman. The fact that the Huffman construction is in terms of probabilities does not matter, since the fact that $p_1 + p_2 + \cdots + p_L = 1$ is never used in the construction or its validity proof. Let us demonstrate the implied procedure by the following example.

Assume $L = 12$ and $m = 4$; the b_i's are given in nonincreasing order: 9, 8, 8, 7, 6, 6, 6, 5, 5, 4, 3, 3. Since $L \equiv 0 \pmod 3$, according to (4.6) $d = 3$.

Thus, in the first step we merge the last three lists to form a list of length 10 which is now put in the first place (see Fig. 4.4). From there on, we merge each time the four lists of least length. The whole merge procedure is described in the tree shown in Fig. 4.5.

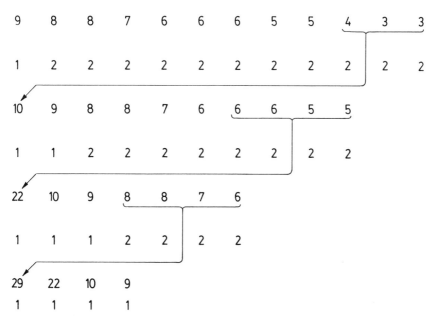

Figure 4.4

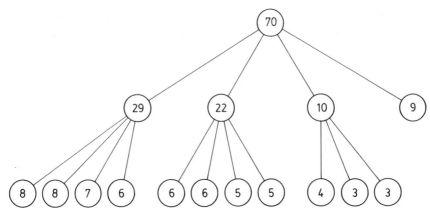

Figure 4.5

4.4 CATALAN NUMBERS

The set of *well-formed sequences of parentheses* is defined by the following recursive definition:

1. The empty sequence is well formed.
2. If *A* and *B* are well-formed sequences, so is *AB* (the concatenation of *A* and *B*).
3. If *A* is well formed, so is (*A*).
4. There are no other well-formed sequences.

For example, (()(())) is well formed; (()))(() is not.

Lemma 4.5: A sequence of (left and right) parentheses is well formed if and only if it contains an even number of parentheses, half of which are left and the other half are right, and as we read the sequence from left to right, the number of right parentheses never exceeds the number of left parentheses.

Proof: First let us prove the "only if" part. Since the construction of every well formed sequence starts with no parentheses (the empty sequence) and each time we add on parentheses (Step 3) there is one left and one right, it is clear that there are n left parentheses and n right parentheses. Now, assume that for every well-formed sequence of m left and m right parentheses, where $m < n$, it is true that as we read it from left to right the number of right parentheses never exceeds the number of left parentheses. If the last step in

the construction of our sequence was 2, then since A is a well-formed sequence, as we read from left to right, as long as we still read A the condition
is satisfied. When we are between A and B, the count of left and right parentheses equalizes. From there on the balance of left and right is safe since B is
well formed and contains less than n parentheses. If the last step in the construction of our sequence was 3, then since A satisfies the condition, so does
(A).

Now, we shall prove the "if" part, again by induction on the number of
parentheses. (Here, as before, the basis of the induction is trivial.) Assume
that the statement holds for all sequences of m left and m right parentheses,
if $m < n$, and we are given a sequence of n left and n right parentheses which
satisfies the condition. Clearly, if after reading $2m$ symbols of it from left to
right the number of left and right parentheses is equal and if $m < n$, this
subsequence, A, by the inductive hypothesis is well formed. Now, the remainder of our sequence, B, must satisfy the condition, too, and again by the
inductive hypothesis is well formed. Thus, by Step 2, AB is well formed. If
there is no such nonempty subsequence A, which leaves a nonempty B, then
as we read from left to right the number of right parentheses, after reading
one symbol and before reading the whole sequence, is strictly less then the
number of left parentheses. Thus, if we delete the first symbol, which is a
"(", and the last, which is a ")", the remainder sequence, A, still satisfies the
condition, and by the inductive hypothesis is well formed. By Step 3 our sequence is well formed too.

Q.E.D.

We shall now show a one-to-one correspondence between the non-
well-formed sequences of n left and n right parentheses, and all sequences
of $n - 1$ left parentheses and $n + 1$ right parentheses.

Let $p_1 p_2 \cdots p_{2n}$ be a sequence of n left and n right parentheses which is not
well formed. By Lemma 4.5, there is a prefix of it which contains more right
parentheses then left. Let j be the least integer such that the number of right
parentheses exceeds the number of left parentheses in the subsequence $p_1 p_2$
$\cdots p_j$. Clearly, the number of right parentheses is then one larger than the
number of left parentheses, or j is not the least index to satisfy the condition.
Now, invert all p_i's for $i > j$ from left parentheses to right parentheses, and
from right parentheses to left parentheses. Clearly, the number of left parentheses is now $n - 1$, and the number of right parentheses is now $n + 1$.

Conversely, given any sequence $p_1 p_2 \cdots p_{2n}$ of $n - 1$ left parentheses and
$n + 1$ right parentheses, let j be the first index such that $p_1 p_2 \cdots p_j$ contains
one right parenthesis more than left parentheses. If we now invert all the
parentheses in the section $p_{j+1} p_{j+2} \cdots p_{2n}$ from left to right and from right to

left, we get a sequence of n left and n right parentheses which is not well formed. This transformation is the inverse of the one of the previous paragraph. Thus, the one-to-one correspondence is established.

The number of sequences of $n - 1$ left and $n + 1$ right parentheses is

$$\binom{2n}{n-1},$$

for we can choose the places for the left parentheses, and the remaining places will have right parentheses. Thus, the number of well-formed sequences of length n is

$$\binom{2n}{n} - \binom{2n}{n-1} = \frac{1}{1+n}\binom{2n}{n}. \tag{4.9}$$

These numbers are called *Catalan numbers*.

An *ordered tree* is a directed tree such that for each internal vertex there is a defined order of its sons. Clearly, every positional tree is ordered, but the converse does not hold: In the case of ordered trees there are no predetermined "potential" sons; only the order of the sons counts, not their position, and there is no limit on the number of sons.

An *ordered forest* is a sequence of ordered trees. We usually draw a forest with all the roots on one horizontal line. The sons of a vertex are drawn from left to right in their given order. For example, the forest shown in Fig. 4.6 consists of three ordered trees whose roots are A, B, and C.

There is a natural correspondence between well-formed sequences of n pairs of parentheses and ordered forests of n vertices. Let us label each leaf with the sequence (). Every vertex whose sons are labeled $w_1, w_2, \ldots, w_s$, is labeled with the concatenation $(w_1 w_2 \cdots w_s)$; clearly, the order of the labels is in the order of the sons. Finally, once the roots are labeled $x_1, x_2, \ldots, x_r$, the sequence which corresponds to the forest is the concatenation $x_1 x_2 \cdots x_n$. For example, the sequence which corresponds to the forest of Fig. 4.6 is $((()())))(()()))((()()))$. The inverse transformation clearly exists and thus the one-to-one correspondence is established. Therefore, the number of ordered forests of n vertices is given by (4.9).

We shall now describe a one-to-one correspondence between ordered forests and positional binary trees. The leftmost root of the forest is the root of the binary tree. The leftmost son of the vertex in the forest is the left son of the vertex in the binary tree. The next brother on the right, or, in the case of a root, the next root on the right is the right son in the binary tree. For example, see Fig. 4.7, where an ordered forest and its corresponding binary tree are

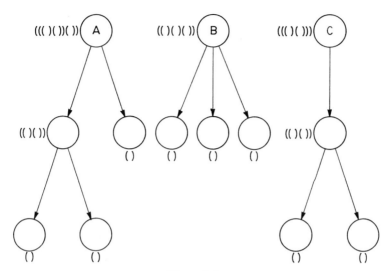

Figure 4.6

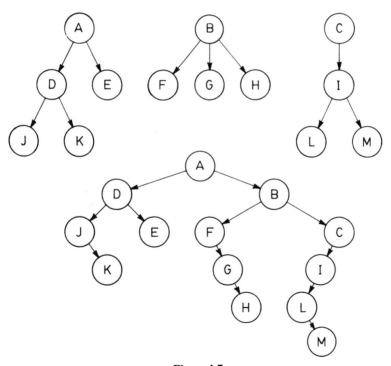

Figure 4.7

drawn. Again, it is clear that this is a one-to-one correspondence and therefore the number of positional binary trees with n vertices is given by (4.9).

There is yet another combinatorial enumeration which is directly related to these.

A *stack* is a storage device which can be described as follows. Suppose that n cars travel on a narrow one-way street where no passing is possible. This leads into a narrow two-way street on which the cars can park or back up to enter another narrow one-way street (see Fig. 4.8). Our problem is to find how may permutations of the cars can be realized from input to output if we assume that the cars enter in the natural order.

The order of operations in the stack is fully described by the sequence of drive-in and drive-out operations. There is no need to specify which car drives in, for it must be the first one on the leading-in present queue; also, the only one which can drive out is the top one in the stack. If we denote a drive-in operation by "(", and a drive-out operation by ")", the whole procedure is described by a well-formed sequence of n pairs of parentheses.

The sequence must be well-formed, by Lemma 4.5, since the number of drive-out operations can never exceed the number of drive-in operations. Also, every well-formed sequence of n pairs of parentheses defines a realizable sequence of operations, since again by Lemma 4.5, a drive-out is never instructed when the stack is empty. Also, different sequences yield different permutations. Thus, the number of permutations on n cars realizable by a stack is given by (4.9).

Let us now consider the problem of finding the number of full binary trees. Denote the number of leaves of a binary tree T by $L(T)$, and the number of internal vertices by $I(T)$. It is easy to prove, by induction on the number of leaves, that $L(T) = I(T) + 1$. Also, if all leaves of T are removed, the resulting tree of $I(T)$ vertices is a positional binary tree T'. Clearly, different T-s will yield different T'-s, and one can reconstruct T from T' by at-

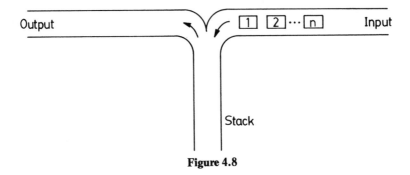

Figure 4.8

taching two leaves to each leaf of T', and one leaf (son) to each vertex which in T' has only one son. Thus, the number of full-binary trees of n vertices is equal to the number of positional binary trees of $(n-1)/2$ vertices. By (4.9) this number is

$$\frac{2}{n+1} \binom{n-1}{\dfrac{n-1}{2}}$$

PROBLEMS

4.1 Prove the following theorem: A position between two letters in a message m over a UD code C is a separation between two code-words if and only if both the prefix and the suffix of m up to this position are messages over C.

4.2 Use the result of Problem 4.1 to construct an efficient algorithm for parsing a message m over a UD code C in order to find the words which compose m. (Hint: Scan m from left to right and mark all the positions such that the prefix m up to them is a message. Repeat from right to left to mark positions which correspond to suffixes which are messages.)

4.3 Test the following codes for UD:

(a) $\{00, 10, 11, 100, 110\}$,
(b) $\{1, 10, 001, 0010, 00000, 100001\}$,
(c) $\{1, 00, 101, 010\}$.

4.4 Construct a prefix binary code, with minimum average code-word length, which consists of 10 words whose probabilities are 0.2, 0.18, 0.12, 0.1, 0.1, 0.08, 0.06, 0.06, 0.06, 0.04. Repeat the construction for $\sigma = 3$ and 4.

4.5 Prove that if a prefix code corresponds to a full positional tree then its characteristic sum is equal to 1.

4.6 Prove that if the word-length vector $(l_1, l_2, \ldots, l_n)$ satisfies the characteristic sum condition and if $l_1 \le l_2 \le \cdots \le l_n$, then there exists a positional tree with n leaves whose levels are the given l_i's and the order of the leaves, from left to right, is as in the vector.

4.7 A code is called *exhaustive* if every word over the alphabet is the beginning of some message over the code. Prove the following:

(a) If a code is prefix and its characteristic sum is 1 then the code is exhaustive.

(b) If a code is UD and exhaustive then it is prefix and its characteristic sum is 1.

4.8 Construct the ordered forest, the positional binary tree and the permutation through a stack which corresponds to the following well-formed sequence of 10 pairs of parentheses:

$$(()(()()))((()()())).$$

4.9 A direct method for computing the number of positional binary trees of n vertices through the use of a generating function goes as follows: Let b_n be the number of trees of n vertices. Define $b_0 = 1$ and define the function

$$B(x) = b_0 + b_1 x + b_2 x^2 + \cdots.$$

(a) Prove that $b_n = b_0 \cdot b_{n-1} + \cdots + b_{n-1} \cdot b_0$.

(b) Prove that $xB^2(x) - B(x) + 1 = 0$.

(c) Use the formula

$$(1 + a)^{1/2} = 1 + \frac{1/2}{1!} a + \frac{1/2(1/2 - 1)}{2!} a^2 + \frac{1/2(1/2 - 1)(1/2 - 2)}{3!} a^3 + \cdots$$

to prove that

$$b_n = \frac{1}{n + 1} \binom{2n}{n}.$$

REFERENCES

[1] Sardinas, A. A., and Patterson, G. W., "A Necessary and Sufficient Condition for the Unique Decomposition of Coded Messages", IRE Convention Record, Part 8, 1953, pp. 104–108.

[2] Gallager, R. G., *Information Theory and Reliable Communication*, John Wiley, 1968. Problem 3.4, page 512.

[3] Levenshtein, V. I., "Certain Properties of Code Systems", Dokl. Akad. Nauk, SSSR, Vol. 140, No. 6, Oct. 1961, pp. 1274-1277. English translation: Soviet Physics, "Doklady", Vol. 6, April 1962, pp. 858-860.

[4] Even, S., "Test for Unique Decipherability", IEEE Trans. on Infor. Th., Vol. IT-9, No. 2, April 1963, pp. 109-112.

[5] Levenshtein, V. I., "Self-Adaptive Automata for Coding Messages", Dokl. Akad. Nauk, SSSR, Vol. 140, Dec. 1961, pp. 1320-1323. English translation: Soviet Physics, "Doklady", Vol. 6, June 1962, pp. 1042-1045.

[6] Markov, Al. A., "On Alphabet Coding", Dokl. Akad. Nauk, SSSR, Vol. 139, July 1961, pp. 560-561. English translation: Soviet Physics, "Doklady", Vol. 6, Jan. 1962, pp. 553-554.

[7] Even, S., "Test for Synchronizability of Finite Automata and Variable Length Codes", IEEE Trans. on Infor. Th., Vol. IT-10, No. 3, July 1964, pp. 185-189.

[8] McMillan, B., "Two Inequalities Implied by Unique Decipherability", IRE Tran. on Infor. Th., Vol. IT-2, 1956, pp. 115-116.

[9] Karush, J., "A Simple Proof of an Inequality of McMillan", IRE Trans. On Infor. Th., Vol. IT-7, 1961, page 118.

[10] Kraft, L. G., "A Device for Quantizing, Grouping and Coding Amplitude Modulated Pulses", M.S. Thesis, Dept. of E.E., M.I.T.

[11] Huffman, D. A., "A Method for the Construction of Minimum Redundancy Codes", Proc. IRE, Vol. 40, No. 10, 1952, pp. 1098-1101.

[12] Perl, Y., Garey, M. R. and Even, S., "Efficient Generation of Optimal Prefix Code: Equiprobable Words Using Unequal Cost Letters", J.ACM, Vol. 22, No. 2, April 1975, pp. 202-214.

[13] Itai, A., "Optimal Alphabetic Trees", SIAM J. Comput., Vol. 5, No. 1, March 1976, pp. 9-18.

[14] Knuth, D. E., *The Art of Computer Programming, Vol. 3: Sorting and Searching,* Addison-Wesley, 1973.

[15] Burge, W. H., "Sorting, Trees, and Measures of Order", *Infor. and Control,* Vol. 1, 1958, pp. 181-197.

Chapter 5

MAXIMUM FLOW IN A NETWORK

5.1 THE FORD AND FULKERSON ALGORITHM

A *network* consists of the following data:

(1) A finite digraph $G(V, E)$ with no self-loops and no parallel edges.*
(2) Two vertices s and t are specified; s is called the *source* and t, the *sink*.**
(3) Each edge $e \in E$ is assigned a non-negative number $c(e)$ called the *capacity* of e.

A *flow function* f is an assignment of a real number $f(e)$ to each edge e, such that the following two conditions hold:

(C1) For every edge $e \in E$, $0 \leq f(e) \leq c(e)$.
(C2) Let $\alpha(v)$ and $\beta(v)$ be the sets of edges incoming to vertex v and outgoing from v, respectively. For every $v \in V - \{s, t\}$

$$0 = \sum_{e \in \alpha(v)} f(e) - \sum_{e \in \beta(v)} f(e). \qquad (5.1)$$

The *total flow* F of f is defined by

$$F = \sum_{e \in \alpha(t)} f(e) - \sum_{e \in \beta(t)} f(e). \qquad (5.2)$$

---

*The exclusion of self-loops and parallel edges is not essential. It will shortly become evident that no generality is lost; the flow in a self-loop gains nothing, and a set of parallel edges can be replaced by one whose capacity is the sum of their capacities. This condition ensures that $|E| \leq |V| \cdot (|V| - 1)$.

**The choice of s or t is completely arbitrary. There is no requirement that s is a graphical source; i.e. has no incoming edges, or that t is a graphical sink; i.e. has no outgoing edges. The edges entering s or leaving t are actually redundant and have no effect on our problem, but we allow them since the choice of s and t may vary, while we leave the other data unchanged.

Namely, F is the net sum of flow into the sink. Our problem is to find an f for which the total flow is maximum.

Let S be a subset of vertices such that $s \in S$ and $t \notin S$. $\bar{S}$ is the complement of S, i.e. $\bar{S} = V - S$. Let $(S; \bar{S})$ be the set of edges of G whose start-vertex is in S and end-vertex is in $\bar{S}$. The set $(\bar{S}; S)$ is defined similarly. The set of edges connecting vertices of S with $\bar{S}$ (in both directions) is called the *cut* defined by S.

By definition, the total flow F is measured at the sink. Our purpose is to show that F can be measured at any cut.

Lemma 5.1: For every S

$$F = \sum_{e \in (S;\bar{S})} f(e) - \sum_{e \in (\bar{S};S)} f(e). \tag{5.3}$$

Proof. Let us sum up equation (5.2) with all the equations (5.1) for $v \in \bar{S} - \{t\}$. The resulting equation has F on the left hand side. In order to see what happens on the right hand side, consider an edge $x \xrightarrow{e} y$. If both x and y belong to S then $f(e)$ does not appear on the r.h.s. at all, in agreement with (5.3). If both x and y belong to $\bar{S}$ then $f(e)$ appears on the r.h.s. once positively, in the equation for y, and once negatively, in the equation for x. Thus, in the summation it is canceled out, again in agreement with (5.3). If $x \in S$ and $y \in \bar{S}$ then $f(e)$ appears on the r.h.s. of the equation for y, positively, and in no other equation we use, and indeed $e \in (S; \bar{S})$, and again we have agreement with (5.3). Finally, if $x \in \bar{S}$ and $y \in S$, $f(e)$ appears negatively on the r.h.s. of the equation for x, and again this agrees with (5.3) since $e \in (\bar{S}; S)$.
Q.E.D.

Let us denote by $c(S)$ the *capacity of the cut* determined by S which is defined as follows:

$$c(S) = \sum_{e \in (S;\bar{S})} c(e). \tag{5.4}$$

Lemma 5.2: For every flow function f, with total flow F, and every S,

$$F \leq c(S). \tag{5.5}$$

Proof: By Lemma 5.1

$$F = \sum_{e \in (S;\bar{S})} f(e) - \sum_{e \in (\bar{S};S)} f(e).$$

By $C1$, $0 \leq f(e) \leq c(e)$ for every $e \in E$. Thus,

$$F \leq \sum_{e \in (S;\bar{S})} c(e) - 0 = c(S).$$

Q.E.D.

A very important corollary of Lemma 5.2, which allows us to detect that a given total flow F is maximum, and a given cut, defined by S, is minimum is the following:

Corollary 5.1: If F and S satisfy (5.5) by equality then F is maximum and the cut defined by S is of minimum capacity.

Ford and Fulkerson [1] suggested the use of augmenting paths to change a given flow function in order to increase the total flow. An *augmenting path* is a simple path from s to t, which is not necessarily directed, but it can be used to advance flow from s to t. If on this path, e points in the direction from s to t, then in order to be able to push flow through it, $f(e)$ must be less than $c(e)$. If e points in the opposite direction, then in order to be able to push through it additional flow from s to t, we must be able to cancel some of its flow. Therefore, $f(e) > 0$ must hold.

In attempt to find an augmenting path for a given flow, a labeling procedure is used. We label s. Then, every vertex v, for which we can find an augmenting path from s to v, is labeled. If t is labeled than an augmenting path has been found. This path is used to increase the total flow, and the procedure is repeated.

A *forward labeling* of vertex v by the edge $u \xrightarrow{e} v$ is applicable if

(1) u is labeled and v is not;
(2) $c(e) > f(e)$.

The label that v gets is $'e'$. If e is used for forward labeling we define $\Delta(e) = c(e) - f(e)$.

A *backward labeling* of vertex v by the edge $u \xleftarrow{e} v$ is applicable if

(1) u is labeled and v is not;
(2) $f(e) > 0$.

The label that v gets is $'e'$. In this case we define $\Delta(e) = f(e)$.
The Ford and Fulkerson algorithm is as follows:

(1) Assign some legal initial flow f to the edges; an assignment $f(e) = 0$ to every edge e will do.

(2) Mark s "labeled" and all other vertices "unlabeled".

(3) Search for a vertex v which can be labeled by either a forward or backward labeling. If none exists, halt; the present flow is maximum. If such a vertex v exists, label it $'e'$, where e is the edge through which the labeling is possible. If $v = t$, go to Step (4); otherwise, repeat Step (3).

(4) Starting from t and by the use of the labels, backtrack the path through which the labeling reached t from s. Let this path be $s = v_0 \overset{e_1}{\rightarrow} v_1 \overset{e_2}{\rightarrow} v_2 \overset{e_3}{\rightarrow} \cdots - v_{l-1} \overset{el}{\rightarrow} v_l = t$. (The directions of the edges are not shown, since each may be in either direction.) Let $\Delta = \text{Min}_{1 \leq i \leq l} \Delta(e_i)$. If e_i is forward, i.e. $v_{i-1} \overset{e_i}{\rightarrow} v_i$, then $f(e_i) \leftarrow f(e_i) + \Delta$. If e_i is backward, i.e. $v_{i-1} \overset{e_i}{\leftarrow} v_i$, then $f(e_i) \leftarrow f(e_i) - \Delta$.

(5) Go to Step (2)

(Note that if the initial flow on the edges entering s is zero, it will never change. This is also true for the edges leaving t.)

As an example, consider the network shown in Fig. 5.1. Next to each edge e we write $c(e), f(e)$ in this order. We assume a zero initial flow everywhere. A first wave of label propagation might be as follows: s is labeled; e_2 used to label c; e_6 used to label d; e_4 used to label a; e_3 used to label b; and finally, e_7 used to label t. The path is $s \overset{e_2}{\rightarrow} c \overset{e_6}{\rightarrow} d \overset{e_4}{\rightarrow} a \overset{e_3}{\rightarrow} b \overset{e_7}{\rightarrow} t$. $\Delta = 4$, and the new flow is shown in Fig. 5.2.

The next augmenting path may be

$$ s \overset{e_1}{\rightarrow} a \overset{e_3}{\rightarrow} b \overset{e_5}{\rightarrow} c \overset{e_6}{\rightarrow} d \overset{e_8}{\rightarrow} t. $$

Now, $\Delta = 3$ and the flow is as in Fig. 5.3.

The next augmenting path may be $s \overset{e_1}{\rightarrow} a \overset{e_3}{\rightarrow} b \overset{e_7}{\rightarrow} t$. Now, $\Delta = 3$ and the new flow is as in Fig. 5.4. Now, the labeling can proceed as follows: s is la-

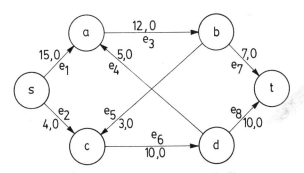

Figure 5.1

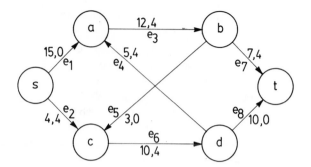

Figure 5.2

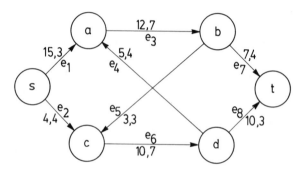

Figure 5.3

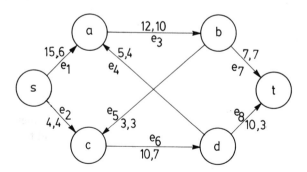

Figure 5.4

beled; e_1 is used to label a; e_3 used to label b; (so far we have not used backward labeling, but this next step is forced) e_4 is used to label d; e_8 is used to label t. The path we backtrack is $s \xrightarrow{e_1} a \xleftarrow{e_4} d \xrightarrow{e_8} t$. Now, $\Delta(e_1) = 9$, $\Delta(e_4) = 4$ and $\Delta(e_8) = 7$. Thus, $\Delta = 4$. The new flow is shown in Fig. 5.5. The next wave of label propagation is as follows:

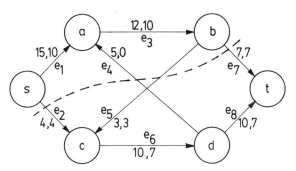

Figure 5.5

s is labeled, e_1 is used to label a, e_3 used to label b. No more labeling is possible and the algorithm halts.

It is easy to see that the flow produced by the algorithm remains legal throughout. The definition of $\Delta(e)$ and Δ guarantees that forward edges will not be overflowed, i.e., $f(e) \leq c(e)$, and that backward edges will not be underflowed, i.e., $f(e) \geq 0$. Also, since Δ is pushed from s to t on a path, the incoming flow will remain equal to the outgoing flow in every vertex $v \in V - \{s, t\}$.

Assuming the algorithm halts, the last labeling process has not reached t. Let S be the set of vertices labeled in the last wave. (In our example $S = \{s, a, b\}$,) If an edge $x \xrightarrow{e} y$ belongs to $(S; \bar{S})$ then it must be saturated, i.e., $f(e) = c(e)$, or a forward labeling could use it. Also, if e belongs to $(\bar{S}; S)$ then it follows that $f(e) = 0$, or a backward labeling could use it. By Lemma 5.1 we have

$$F = \sum_{e \in (S;\bar{S})} f(e) - \sum_{e \in (\bar{S};S)} f(e) = \sum_{e \in (S;\bar{S})} c(e) = c(S).$$

Now, by Corollary 5.1, F is a maximum total flow and S defines a minimum cut. In our example, $(S; \bar{S}) = \{e_2, e_5, e_7\}$, $(\bar{S}; S) = \{e_4\}$ and the value of F is 14.

The question of whether the algorithm will always halt remains to be discussed. Note first a very important property of the Ford and Fulkerson algorithm: If the initial flow is integral, for example, zero everywhere, and if all the capacities are integers, then the algorithm never introduces fractions. The algorithm adds and subtracts, but it never divides. Also, if t is labeled, the augmenting path is used to increase the total flow by at least one unit. Since there is an upper bound on the total flow (any cut), the process must terminate.

Ford and Fulkerson showed that their algorithm may fail, if the capacities are allowed to be irrational numbers. Their counterexample (Reference 1, p. 21) displays an infinite sequence of flow augmentations. The flow converges (in infinitely many steps) to a value which is one fourth of the maximum total flow. We shall not bring their example here; it is fairly complex and as the reader will shortly discover, it is not as important any more.

One could have argued that for all practical purposes, we may assume that the algorithm is sure to halt. This follows from the fact that our computations are usually through a fixed radix (decimal, binary, and so on) number representation with a bound on the number of digits used; in other words, all figures are multiples of a fixed quantum and the termination proof works here as it does for integers. However, a simple example shows the weakness of this argument. Consider the network shown in Fig. 5.6. Assume that M is a very large integer. If the algorithm starts with $f(e) = 0$ for all e, and alternatively uses $s - a - b - t$ and $s - b - a - t$ as augmenting paths, it will take $2M$ augmentations before $F = 2M$ is achieved.

Edmonds and Karp [2] were first to overcome this problem. They showed that if one uses breadth-first search (BFS) in the labeling algorithm and always uses a shortest augmenting path, the algorithm will terminate in $O(|V|^3|E|)$ steps, regardless of the capacities. (Here, of course, we assume

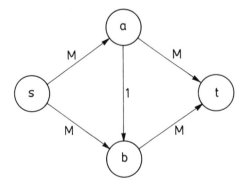

Figure 5.6

that our computer can handle, in one step, any real number.) In the next section we shall present the more advanced work of Dinic [3]; his algorithm has time complexity $O(|V|^2|E|)$. Karzanov [4] and Cherkassky [5] have reduced it to $O(|V|^3)$ and $O(|V|^2|E|^{1/2})$, respectively. This algorithms are fairly complex and will not be described. A recent algorithm of Malhotra, Pramodh Kumar and Maheshwari [6] has the same time complexity as Karzanov's and is much simpler; it will be described in the next section.

The existence of these algorithms assures that, if one proceeds according to a proper strategy in the labeling procedure, the algorithm is guaranteed to halt. When it does, the total flow is maximum, and the cut indicated is minimum, thus providing the *max-flow min-cut* theorem:

Theorem 5.1: Every network has a maximum total flow which is equal to the capacity of a cut for which the capacity is minimum.

5.2   THE DINIC ALGORITHM

As in the Ford and Fulkerson algorithm, the Dinic algorithm starts with some legal flow function f and improves it. When no improvement is possible the algorithm halts, and the total flow is maximum.

If presently an edge $u \xrightarrow{e} v$ has flow $f(e)$ then we say that e is *useful* from u to v if one of the following two conditions holds:

(1) $u \xrightarrow{e} v$ and $f(e) < c(e)$.
(2) $u \xleftarrow{e} v$ and $f(e) > 0$.

The *layered network* of $G(V, E)$ with a flow f is defined by the following algorithm:

(1) $V_0 \leftarrow \{s\}$, $i \leftarrow 0$.
(2) Construct $T \leftarrow \{v \mid v \notin V_j \text{ for } j \leq i \text{ and there is a useful edge from a vertex of } V_i \text{ to } v\}$.
(3) If T is empty, the present total flow F is maximum, halt
(4) If T contains t then $l \leftarrow i + 1$, $V_l \leftarrow \{t\}$ and halt.
(5) Let $V_{i+1} \leftarrow T$, increment i and return to Step (2).

For every $1 \leq i \leq 1$, let E_i be the set of edges useful from a vertex of V_{i-1} to a vertex of V_i. The sets V_i are called *layers*.

The construction of the layered network investigates each edge at most twice; once in each direction. Thus, the time complexity of this algorithm is $O(|E|)$.

Lemma 5.3: If the construction of the layered network terminates in Step (3) then the present total flow, F, is indeed maximum.

Proof: The proof here is very similar to the one in the Ford and Fulkerson algorithm: Let S be the union of $V_0, V_1, \ldots, V_i$. Every edge $u \overset{e}{\rightarrow} v$ in $(S; \bar{S})$ is saturated, i.e. $f(e) = c(e)$, or else e is useful from u to v and T is not empty. Also, every edge $u \overset{e}{\leftarrow} v$ is $(\bar{S}; S)$ has $f(e) = 0$, or again e is useful from u to v, etc. Thus, by Lemma 5.1,

$$F = \sum_{e \in (S;\bar{S})} f(e) - \sum_{e \in (\bar{S};S)} f(e) = \sum_{e \in (S;\bar{S})} c(e) - 0 = c(S).$$

By Corollary 5.1, F is maximum.

Q.E.D.

For every edge e in E_j let $\tilde{c}(e)$ be defined as follows:

(i) If $u \in V_{j-1}$, $v \in V_j$ and $u \overset{e}{\rightarrow} v$ then $\tilde{c}(e) = c(e) - f(e)$.
(ii) If $u \in V_{j-1}$, $v \in V_j$ and $u \overset{e}{\leftarrow} v$ then $\tilde{c}(e) = f(e)$.

We now consider all edges of E_j to be directed from V_{j-1} to V_j, even if in $G(V, E)$ they may have the opposite direction (in case (ii)). Also, the initial flow in the new network is $\tilde{f}(e) = 0$ everywhere. We seek a maximal flow $\tilde{f}$ in the layered network; by a *maximal flow $\tilde{f}$* we mean that $\tilde{f}$ satisfies the condition that for every path $s \overset{e_1}{\rightarrow} v_1 \overset{e_2}{\rightarrow} v_2 — \cdots v_{l-1} \overset{e_l}{\rightarrow} t$, where $v_j \in V_j$ and $e_j \in E_j$, there is at least one edge e_j such that $\tilde{f}(e_j) = \tilde{c}(e_j)$. Clearly, a maximal flow is not necessarily maximum as the example of Figure 5.7 shows: If for all edges $\tilde{c} = 1$ and we push one unit flow from s to t via a and d then the resulting flow is maximal in spite of the fact that the total flow is 1 while a total flow of 2 is possible.

Later we shall describe how one can find a maximal flow function $\tilde{f}$ efficiently. For now, let us assume that such a flow function has been found and its total value is $\tilde{F}$. The flow f in $G(V, E)$ is changed into f' as follows:

(i) If $u \overset{e}{\rightarrow} v$, $u \in V_{j-1}$ and $v \in V_j$ then $f'(e) = f(e) + \tilde{f}(e)$.
(ii) If $u \overset{e}{\leftarrow} v$, $u \in V_{j-1}$ and $v \in V_j$ then $f'(e) = f(e) - \tilde{f}(e)$.

It is easy to see that the new flow f' satisfies both $C1$ (due to the choice of $\tilde{c}$) and $C2$ (because it is the superposition of two flows which satisfy $C2$). Clearly $F' = F + \tilde{F} > F$.

Let us call the part of the algorithm which starts with f, finds its layered network, finds a maximal flow $\tilde{f}$ in it and improves the flow in the original

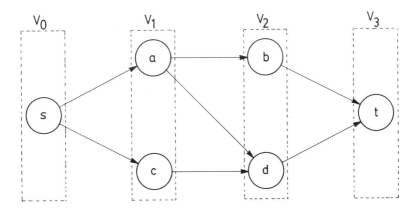

Figure 5.7

network to become f' — a *phase*. We want to show that the number of phases is bounded by $|V|$. For this purpose we shall prove that the length of the layered network increases from phase to phase; by *length* is meant the index of the last layer, which we called l in Step (4) of the layered network algorithm. Thus, l_k denotes the length of the layered network of the kth phase.

Lemma 5.4: If the $(k + 1)$st phase is not the last then $l_{k+1} > l_k$.

Proof: There is a path of length $k + 1$ in the $(k + 1)$st layered network which starts with s and ends with t:

$$s \xrightarrow{e_1} v_1 \xrightarrow{e_2} \cdots v_{l_{k+1}-1} \xrightarrow{e_{l_{k+1}}} t.$$

First, let us assume that all the vertices of the path appear in the k-th layered network. Let V_j be the jth layer of the kth layered network. We claim that if $v_a \in V_b$ then $a \geq b$. This is proved by induction on a. For $a = 0$, ($v_0 = s$) the claim is obviously true. Now, assume $v_{a+1} \in V_c$. If $c \leq b + 1$ the inductive step is trivial. But if $c > b + 1$ then the edge e_{a+1} has not been used in the kth phase since it is not even in the kth layered network, in which only edges between adjacent layers appear. If e_{a+1} has not been used and is useful from v_a to v_{a+1} in the beginning of phase $k + 1$, then it was useful from v_a to v_{a+1} in the beginning of phase k. Thus, v_{a+1} cannot belong to V_c (by the algorithm). Now, in particular, $t = v_{l_k+1}$ and $t \in V_{l_k}$. Therefore, $l_{k+1} \geq l_k$. Also, equality cannot hold, because in this case the whole path is in the kth layered network, and if all its edges are still useful in the beginning of phase $k + 1$ then the $\tilde{f}$ of phase k was not maximal.

If not all the vertices of the path appear in the kth layered network then let $v_a \xrightarrow{e_{a+1}} v_{a+1}$ be the first edge such that for some b $v_a \in V_b$ but v_{a+1} is not in the kth layered network. Thus, e_{a+1} was not used in phase k. Since it is useful in the beginning of phase $k + 1$, it was also useful in the beginning of phase k. The only possible reason for v_{a+1} not to belong to V_{b+1} is that $b + 1 = l_k$. By the argument of the previous paragraph $a \geq b$. Thus $a + 1 \geq l_k$, and therefore $l_{k+1} > l_k$.

Q.E.D.

Corollary 5.2: The number of phases is less than or equal to $|V| - 1$.

Proof: Since $l \leq |V| - 1$, Lemma 5.4 implies the corollary.

Q.E.D.

The remaining task is to describe an efficient algorithm to construct a maximal flow in a layered network.

First, let us show Dinic's method.

We assume that $\tilde{N}$ is a layered network, and for every edge e in $\tilde{N}$ $\tilde{c}(e) > 0$.

(1) For every e in $\tilde{N}$, mark e "unblocked" and let $\tilde{f}(e) \leftarrow 0$.
(2) $v \leftarrow s$ and empty the stack S.
(3) If there is no unblocked edge $v - u$, with u in the next layer, then (v is a *dead-end* and) perform the following operations:
    (3.1) If $s = v$, halt; the present $\tilde{f}$ is maximal.
    (3.2) Delete the top-most edge $u \xrightarrow{e} v$ from S.
    (3.3) Mark e "blocked" and let $v \leftarrow u$.
    (3.3) Repeat Step (3).
(4) Choose an unblocked edge $v \xrightarrow{e} u$, with u in the next layer. Put e in S and let $v \leftarrow u$. If $v \neq t$ then go to Step (3).
(5) The edges on S form an augmenting path:
    $s \xrightarrow{e_1} v_1 \xrightarrow{e_2} v_2 \xrightarrow{e_3} \cdots v_{l-1} \xrightarrow{e_l} t$. Perform the following operations:
    (5.1) $\Delta \leftarrow \text{Min}_{1 \leq i \leq l} (\tilde{c}(e_i) - \tilde{f}(e_i))$.
    (5.2) For every $1 \leq i \leq l$, $\tilde{f}(e_i) \leftarrow \tilde{f}(e_i) + \Delta$ and if $\tilde{f}(e_i) = \tilde{c}(e_i)$ then mark e_i "blocked".
    (5.3) Go to Step (2).

It is easy to see that an edge is declared "blocked" only if no additional augmenting path (of length l) can use it. Thus, when the algorithm halts (in Step (3.1)) the resulting flow $\tilde{f}$ is maximal in $\tilde{N}$. Also, the number of edges scanned, in between two declarations of edge blocking, is at most l, and $l \leq |V| - 1$. Since the number of edges in $\tilde{N}$ is at most $|E|$ and since no blocked edge becomes unblocked, the number of edge scannings is bounded by $|V| \cdot |E|$. Thus, the algorithm for finding a maximal flow in $\tilde{N}$ is $O(|V| \cdot |E|)$, yielding a bound $O(|V|^2|E|)$ for the whole algorithm.

Next, let us describe the MPM algorithm [6] for finding a maximal flow in $\tilde{N}$. This algorithm (following Karzanov) does not use augmenting paths. Let $\tilde{V}$ be the set of vertices in $\tilde{N}$. Clearly, $\tilde{V} = V_0 \cup V_1 \cup \cdots \cup V_l$. For every vertex $v \in \tilde{V}$, let $\tilde{\alpha}(v)(\tilde{\beta}(v))$ be the set of edges $u \overset{e}{\longrightarrow} v$ ($v \overset{e}{\longrightarrow} u$) such that u belongs to the previous (next) layer. Let $\tilde{\alpha}(v) = \{e_{v1}, e_{v2}, \ldots, e_{vp_v}\}$ and $\tilde{\beta}(v) = \{e'_{v1}, e'_{v2}, \ldots, e'_{vq_v}\}$.

(1) For every $v \in \tilde{V}$, $IP(v) \leftarrow \sum\limits_{i=1}^{p_v} \tilde{c}(e_{vi})$, $OP(v) \leftarrow \sum\limits_{i=1}^{q_v} \tilde{c}(e'_{vi})$, $m_v \leftarrow 1, m'_v \leftarrow 1$.

(2) For every e in $\tilde{N}$, $\tilde{f}(e) \leftarrow 0$.

(3) Perform the following operations:

    (3.1) $P(s) \leftarrow OP(s)$, $P(t) \leftarrow IP(v)$.

    (3.2) For $v \in \tilde{V} - \{s, t\}$, $P(v) \leftarrow \text{Min}\{IP(v), OP(v)\}$.

    (3.3) Find a vertex v for which $P(v)$ is minimum.

(4) If $P(v) = 0$, perform the following operations:

    (4.1) If $v = s$ or t, halt; the present f is maximal.

    (4.2) For every $m_v \leq m \leq p_v$, where $u \overset{e_{vm}}{\longrightarrow} v$, if $u \in \tilde{V}$ then $OP(u) \leftarrow OP(u) - (\tilde{c}(e_{vm}) - \tilde{f}(e_{vm}))$.

    (4.3) For every $m_v' \leq m' \leq q_v$, where $v \overset{e_{vm'}}{\longrightarrow} u$, if $u \in \tilde{V}$ then $IP(u) \leftarrow IP(u) - (\tilde{c}(e'_{vm'}) - \tilde{f}(e'_{vm'}))$.

    (4.4) $\tilde{V} \leftarrow \tilde{V} - \{v\}$ and go to (3).

(5) Find i for which $v \in V_i$. Let $j \leftarrow i$, $k \leftarrow i$. Also $OF(v) \leftarrow P(v)$, $IF(v) \leftarrow P(v)$ and for every $u \in \tilde{V} - \{v\}$ let $OF(u) \leftarrow 0$ and $IF(u) \leftarrow 0$.

(6) If $k = l$ (the pushing from v to t is complete) go to (9).

(7) Assume $V_k = \{v_1, v_2, \ldots, v_{k'}\}$.

    (7.1) $r \leftarrow 1$.

    (7.2) $u \leftarrow v_r$.

    (7.3) If $OF(u) = 0$ then go to (7.5).

    (7.4) $(OF(u) > 0$. Starting with $e'_{um'_u}$, we push the excess supply towards t.)

        (7.4.1) $e \leftarrow e'_{um'_u}$ and assume $u \overset{e}{\longrightarrow} w$.

(7.4.2) If $\tilde{f}(e) = \tilde{c}(e)$ or $w \notin \tilde{V}$ then go to (7.4.5).

(7.4.3) If $OF(u) \leq \tilde{c}(e) - \tilde{f}(e)$ then perform the following operations:

$$\tilde{f}(e) \leftarrow \tilde{f}(e) + OF(u),$$
$$OF(w) \leftarrow OF(w) + OF(u),$$
$$OP(u) \leftarrow OP(u) - OF(u),$$
$$IP(w) \leftarrow IP(w) - OF(u),$$

and go to (7.5).

(7.4.4) $(OF(u) > \tilde{c}(e) - \tilde{f}(e))$
$$OF(u) \leftarrow OF(u) - (\tilde{c}(e) - \tilde{f}(e)),$$
$$OF(w) \leftarrow OF(w) + (\tilde{c}(e) - \tilde{f}(e)),$$
$$OP(u) \leftarrow OP(u) - (\tilde{c}(e) - \tilde{f}(e)),$$
$$IP(w) \leftarrow IP(w) - (\tilde{c}(e) - \tilde{f}(e)),$$
$$\tilde{f}(e) \leftarrow \tilde{c}(e).$$

(7.4.5) $m'_u \leftarrow m'_u + 1$ and go to (7.4.1).

(7.5) If $r = k'$, go to (8).

(7.6) $r \leftarrow r + 1$ and go to (7.2).

(8) $k \leftarrow k + 1$ and go to (6).

(9) If $j = 1$ (the pulling from s to v is complete too) then $\tilde{V} \leftarrow \tilde{V} - \{v\}$ and go to (3).

(10) Assume $V_j = \{v_1, v_2, \ldots, v_{j'}\}$.

(10.1) $r \leftarrow 1$.

(10.2) $u \leftarrow v_r$.

(10.3) If $IF(u) = 0$ then go to (10.5).

(10.4) $(IF(u) > 0$. Starting with e_{um_u}, we pull the excess demand into u.)

(10.4.1) $e \leftarrow e_{um_u}$ and assume $w \xrightarrow{e} u$.

(10.4.2) If $\tilde{f}(e) = \tilde{c}(e)$ or $w \notin \tilde{V}$ then go to (10.4.5).

(10.4.3) If $IF(u) \leq \tilde{c}(e) - \tilde{f}(e)$ then perform the following operations:

$$\tilde{f}(e) \leftarrow \tilde{f}(e) + IF(u)$$
$$IF(w) \leftarrow IF(w) + IF(u)$$
$$IP(u) \leftarrow IP(u) - IF(u),$$
$$OP(w) \leftarrow OP(w) - IF(u)$$

and go to (10.5).

(10.4.4) $(IF(u) > \tilde{c}(e) - \tilde{f}(e))$
$$IF(u) \leftarrow IF(u) - (\tilde{c}(e) - \tilde{f}(e)),$$
$$IF(w) \leftarrow IF(w) + (\tilde{c}(e) - \tilde{f}(e)),$$
$$IP(u) \leftarrow IP(u) - (\tilde{c}(e) - \tilde{f}(e)),$$
$$OP(w) \leftarrow OP(w) - (\tilde{c}(e) - \tilde{f}(e)),$$
$$\tilde{f}(e) \leftarrow \tilde{c}(e).$$

$(10.4.5)$ $m_u \leftarrow m_u + 1$ and go to $(10.4.1)$
(10.5) If $r = j'$, go to (11).
(10.6) $r \leftarrow r + 1$ and go to (10.2).
(11) $j \leftarrow j - 1$ and go to (9).

First, we compute for every vertex v its *in-potential IP(v)*, which is a local upper bound on the flow which can enter v. Similarly, the *out-potential OP(v)*, is computed. For every vertex $v \neq s$ or t, the *potential, P(v)*, is the minimum of $IP(v)$, and $OP(v)$. For s, $P(s) = OP(s)$ and for t, $P(t) = IP(t)$. Next, we find v for which $P(v)$ is minimum.

The main idea is that we can easily find a flow of $P(v)$ units which goes from s to t via v. We use the edges of $\beta(v)$, one by one, saturating them as long as the excess supply lasts, and pushing through them flow to the next layer. For each of the vertices on higher layers we repeat the same process, until all the $P(v)$ units reach t. This is done in Steps (6) to (8). We can never get stuck with too much excess supply in a vertex, since v is of minimum potential. We then do the same while pulling the excess demand, $P(v)$, into v, from the previous layer, and then into it from the layer preceeding it, etc. This is done in Steps (9) to (11). When this is over, we return to (3) to choose a vertex v for which $P(v)$ is now minimum (Step (3)), and repeat the pushing and pulling for it.

Clearly, when edges incident to a vertex are used, their in-potential and out-potential must be updated. Also, variables $IF(v)$ and $OF(v)$ are used to record the excess demand that should be flowed into v, and the excess supply that should be flowed out of v, respectively. If $P(v) = 0$, none of v's incident edges can be used anymore for flowing additional flow from s to t. Thus, the in and out potentials of the adjacent vertices are updated accordingly; this is done in Step (4). If $P(s)$ or $P(t)$ is zero, the flow is maximal, and the algorithm halts (see (4.1)).

Every edge can be saturated once only (in $(7.4.4)$ or $(10.4.4)$). The number of all other uses of edges (in $(7.4.3)$ or $(10.4.3)$) can be bounded as follows:

For every v, when $P(v)$ is minimum and we push and pull for v, for every $u \neq v$ we use at most one outgoing edge without saturating it (in $(7.4.3)$) or one incoming edge (in $(10.4.3)$). Thus, the number of edge-uses is bounded by $E + |V|^2 = O(|V|^2)$. Thus, the complexity of the MPM algorithm for finding maximal flow in a layered network is $O(|V|^2)$ and if we use it, the maximum flow problem is solved in $O(|V|^3)$ time, since the number of phases is bounded by $|V|$.

5.3 NETWORKS WITH UPPER AND LOWER BOUNDS

In the previous sections we have assumed that the flow in the edges is bounded from above but the lower bound on all the edges is zero. The significance of this assumption is that the assignment of $f(e) = 0$, for every edge e, defines a legal flow, and the algorithm for improving the flow can be started without any difficulty.

In this section, in addition to the upper bound, $c(e)$, on the flow through e, we assume that the flow is also bounded from below by $b(e)$. Thus, f must satisfy

$$b(e) \leq f(e) \leq c(e) \tag{5.6}$$

in every edge e. Condition $C2$ remains unchanged.

Thus, our problem of finding a maximum flow is divided into two. First, we want to check whether the given network has legal flows, and if the answer is positive, we want to find one. Second, we want to increase the flow and find a maximum flow.

A simple example of a network which has no legal flow is shown in Figure 5.8. Here next to each edge e we write $b(e)$, $c(e)$.

Figure 5.8

The following method for testing whether a given network has a legal flow function is due to Ford and Fulkerson [1]. In case of a positive answer, a flow function is found.

The original network with graph $G(V, E)$ and bounds $b(e)$ and $c(e)$ is modified as follows:

(1) The new set of vertices, $\bar{V}$, is defined by

$$\bar{V} = \{\bar{s}, \bar{t}\} \cup V.$$

$\bar{s}$ and $\bar{t}$ are new vertices, called the auxiliary source and sink, respectively,

(2) For every $v \in V$ construct an edge $v \overset{e}{\to} \bar{t}$ with an upper bound (capacity)

$$\bar{c}(e) = \sum_{e \in \beta(v)} b(e),$$

where $\beta(v)$ is the set of edges which emanate from v in G. The lower bound is zero.

(3) For every $v \in V$ construct an edge $\bar{s} \overset{e}{\to} v$ with an upper bound

$$\bar{c}(e) = \sum_{e \in \alpha(v)} b(e),$$

where $\alpha(v)$ is the set of edges which enter v in G. The lower bound is zero.

(4) The edges of E remain in the new graph but the bounds change: The lower bounds are all zero and the upper bound $\bar{c}(e)$ of $e \in E$ is defined by $\bar{c}(e) = c(e) - b(e)$.

(5) Construct new edges $s \overset{e}{\to} t$ and $t \overset{e'}{\to} s$ with very high upper bounds $\bar{c}(e)$ and $\bar{c}(e')(= \infty)$ and zero lower bounds.

The resulting auxiliary network has a source $\bar{s}$, a sink $\bar{t}$; s and t are regarded now as regular vertices which have to conform to the conservation rule, i.e. condition $C2$.

Let us demonstrate this construction on the graph shown in Fig. 5.9(a). The auxiliary network is shown in Fig. 5.9(b). The upper bounds $\bar{c}(e)$ are shown next to the edges to which they apply.

Now we can use the Ford and Fulkerson or the Dinic (with or without the MPM improvement) algorithms to find a maximum flow in the auxiliary network.

Theorem 5.2: The original network has a legal flow if and only if the maximum flow of the auxiliary network saturates all the edges which emanate from $\bar{s}$.

Clearly, if all the edges which emanate from $\bar{s}$ are saturated, then so are all the edges which enter $\bar{t}$. This follows from the fact that each $b(e)$, of the original graph, contributes its value to the capacity of one edge emanating from $\bar{s}$ and to the capacity of one edge entering $\bar{t}$. Thus, the sum of capacities of edges emanating from $\bar{s}$ is equal to the sum of capacities of edges entering $\bar{t}$.

Proof: Assume a maximum flow function $\bar{f}$ of the auxiliary network saturates all the edges which emanate from $\bar{s}$. Define the following flow function, for the original network:

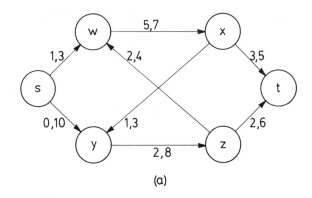

(a)

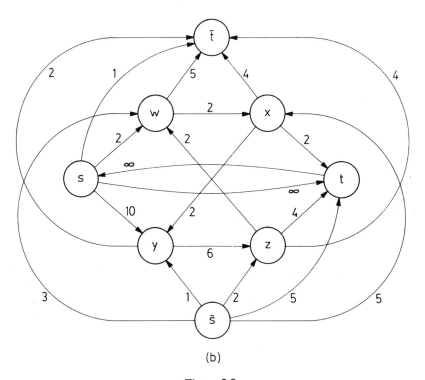

(b)

Figure 5.9

For every $e \in E$

$$f(e) = \bar{f}(e) + b(e). \tag{5.7}$$

Since

$$0 \leq \bar{f}(e) \leq \bar{c}(e) = c(e) - b(e),$$

we have

$$b(e) \leq f(e) \leq c(e),$$

satisfying (5.6).

Now let $v \in V - \{s, t\}$; $\alpha(v)$ is the set of edges which enter v in the original network and $\beta(v)$ is the set of edges which emanate from v in it. Let $\bar{s} \xrightarrow{\sigma} v$ and $v \xrightarrow{\tau} \bar{t}$ be the edges of the auxiliary network, as constructed in parts (3) and (2). Clearly,

$$\sum_{e \in \alpha(v)} \bar{f}(e) + \bar{f}(\sigma) = \sum_{e \in \beta(v)} \bar{f}(e) + \bar{f}(\tau). \tag{5.8}$$

By the assumption

$$\bar{f}(\sigma) = \bar{c}(\sigma) = \sum_{e \in \alpha(v)} b(e)$$

and

$$\bar{f}(\tau) = \bar{c}(\tau) = \sum_{e \in \beta(v)} b(e).$$

Thus

$$\sum_{e \in \alpha(v)} f(e) = \sum_{e \in \beta(v)} f(e). \tag{5.9}$$

This proves that $C2$ is satisfied too, and f is a legal flow function of the original network.

The steps of this proof are reversible, with minor modifications. If f is a legal flow function of the original network, we can define $\bar{f}$ for the auxiliary network by (5.7). Since f satisfies (5.6), by subtracting $b(e)$, we get that $\bar{f}(e)$ satisfies $C1$ in $e \in E$. Now, f satisfies (5.9) for every $v \in V - \{s, t\}$. Let $\bar{f}(\sigma) = \bar{c}(\sigma)$ and $\bar{f}(\tau) = \bar{c}(\tau)$. Now (5.8) is satisfied and therefore condition $C2$ is held while all the edges which emanate from $\bar{s}$ are saturated. Finally, since the net flow which emanates from s is equal to the net flow which enters t, we can make both of them satisfy $C2$ by flowing through the edges of part (5) of the construction, this amount.

Q.E.D.

Let us demonstrate the technique for establishing whether the network has a legal flow, and finding one in the case the answer is positive, on our example (Fig. 5.9). First, we apply the Dinic algorithm on the auxiliary network and end up with the flow, as in Fig. 5.10(a). The maximum flow saturates all the edges which emanate from $\bar{s}$, and we conclude that the original network has a legal flow. We use (5.7) to define a legal flow in the original network; this is shown in Fig. 5.10(b) (next to each edge e we write $b(e)$, $c(e)$, $f(e)$, in this order).

Once a legal flow has been found, we turn to the question of optimizing it. First, let us consider the question of maximizing the total flow.

One can use the Ford and Fulkerson algorithm except that the backward labeling must be redefined as follows:

A *backward labeling* of vertex v by the edge $u \overset{e}{\leftarrow} v$ is applicable if:

(1) u is labeled and v is not;
(2) $f(e) > b(e)$.

The label that v gets is $'e'$. In this case we define $\Delta(e) = f(e) - b(e)$.

We start the algorithm with the known legal flow. With this exception, the algorithm is exactly as described in Section 5.1. The proof that when the algorithm terminates the flow is maximum is similar too. We need to redefine the *capacity of a cut* determined by S as follows:

$$c(S) = \sum_{e \in (S;\bar{S})} c(e) - \sum_{e \in (\bar{S};S)} b(e).$$

It is easy to prove that the statement analogous to Lemma 5.2, still holds; for every flow f with total flow F and every S

$$F \leq c(S). \tag{5.10}$$

Now, the set of labeled vertices S, when the algorithm terminates satisfies (5.10) by equality. Thus, the flow is maximum and the indicated cut is minimum.

The Dinic algorithm can be used too. The only change needed is in the definition of a useful edge, part (ii): $u \overset{e}{\leftarrow} v$ and $f(e) > b(e)$, instead of

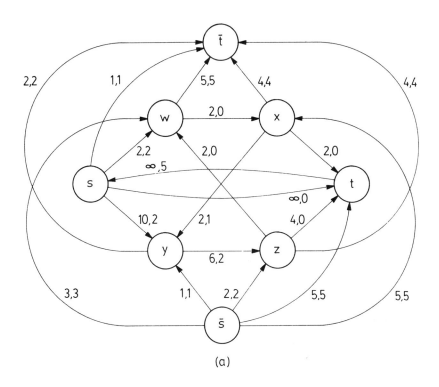

(a)

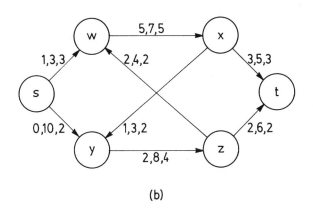

(b)

Figure 5.10

$f(e) > 0$. Also, in the definition of $\bar{c}(e)$, part (ii): If $u \in V_{i-1}$, $v \in V_i$ and $u \xrightarrow{e} v$ then $\bar{c}(e) = f(e) - b(e)$.

Let us demonstrate the maximizing of the flow on our example, by the Dinic algorithm. The layered network of the first phase for the network, with legal flow, of Fig. 5.10(b) is shown in Fig. 5.11(a). The pair $\bar{c}(e)$, $\tilde{f}(e)$ is shown next to each edge. The new flow of the original network is shown in Fig. 5.11(b). The layered network of the second phase is shown in Fig. 5.11(c). The set $S = \{s, y\}$ indicates a minimum cut, and the flow is maximum.

In certain applications, what we want is a *minimum flow*, i.e. a legal flow function f for which the total flow F is minimum. Clearly, a minimum flow from s to t is a maximum flow from t to s. Thus, our techniques solve this problem too, by simply exchanging the roles of s and t. By the max-flow min-cut theorem, the max-flow from t to s, $F(t, s)$ is equal to a min-cut from t to s. Therefore, there exists a $T \subset V$, $t \in T$, $s \notin T$ such that

$$F(t, s) = c(T) = \sum_{e \in (T;\bar{T})} c(e) - \sum_{e \in (\bar{T};T)} b(e).$$

Now $F(t, s) = -F(s, t)$, and if $S = \bar{T}$, then

$$F(s, t) = \sum_{e \in (S;\bar{S})} b(e) - \sum_{e \in (\bar{S};S)} c(e).$$

For the min-flow problem we define the capacity of a cut determined by S, $s \in S$, $t \notin S$ by

$$c(S) = \sum_{e \in (S;\bar{S})} b(e) - \sum_{e \in (\bar{S};S)} c(e).$$

Clearly, every S yields a lower bound, $c(S)$, on the flow $F(s, t)$ and the min-flow is equal to the max-cut.

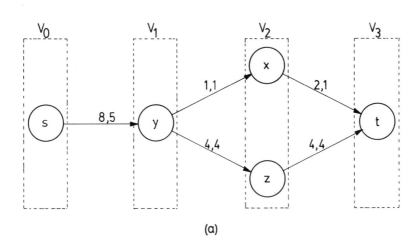

(a)

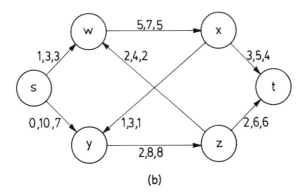

(b)

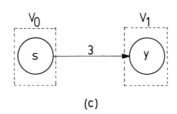

(c)

Figure 5.11

PROBLEMS

5.1 Find a maximum flow in the network shown below. The number next to each edge is its capacity.

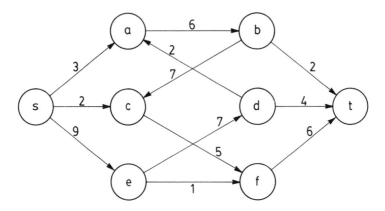

5.2 In the following network, x_1, x_2, x_3 are all sources (of the same commodity). The supply available at x_1 is 5, at x_2 is 10, and at x_3 is 5. The vertices y_1, y_2, y_3 are all sinks. The demand required at y_1 is 5, at y_2 is 10, and at y_3 is 5. Find out whether all the requirements can be met simultaneously. (Hint: One way of solving this type of problem is to introduce an auxiliary source s and a sink t; connect s to x_i through an edge of capacity equal to x_i's supply; connect each y_i to t through an edge of capacity equal to y_i's demand; find a maximum flow in the resulting network and observe if all the demands are met.)

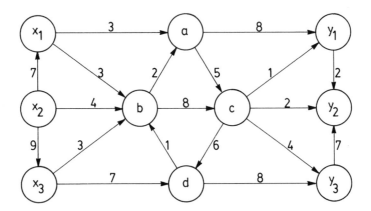

5.3 In the following network, in addition to the capacities of the edges, each vertex other than s and t has an upper bound on the flow that may flow through it. These vertex capacities are written below the vertex labels. Find a maximum flow for this network. (Hint: One way of solving this type of problem is to replace each vertex v by two vertices v' and v'' with an edge $v' \xrightarrow{e} v''$, where $c(e)$ is the upper bound on the flow that may go through v in the original network. All the edges which previously entered v are now entering v', and all the edges which previously emanated from v now emanate from v''.)

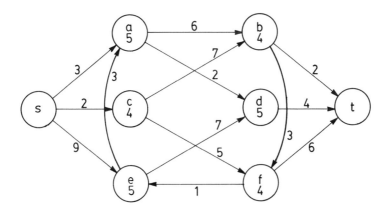

5.4 (a) Describe an alternative labeling procedure, like that of Ford and Fulkerson, for maximizing the flow, except that the labeling starts at t, and if it reaches s an augmenting path is found.
(b) Demonstrate your algorithm on the following network.

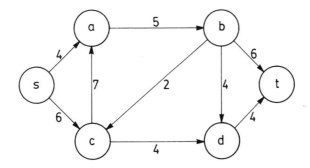

(c) Describe a method of locating an edge which has the property that increasing its capacity increases the maximum flow in the graph.

(Hint: One way of doing this is to use both source-to-sink and sink-to-source labelings.) Demonstrate your method on the graph of (b).
(d) Does an edge like this always exist? Prove your claim.

5.5 Prove that in a network with a nonnegative lower bound $b(e)$ for every edge e but no upper bound ($c(e) = \infty$), there exists a legal flow if and only if for every edge e either e is in a directed circuit or e is in a directed path from s to t or from t to s.

5.6 Find a minimum flow from s to t for the network of Problem 5.1, where all the numbers next to the edges are now assumed to be lower bounds, and there are no upper bounds ($=\infty$).

5.7 The two networks shown have both lower and upper bounds on the flow through the edges. Which of the two networks has no legal flow? Find both a maximum flow and minimum flow if a legal flow exists. If no legal flow exists display a set of vertices which neither includes the source, nor the sink, and is either required to "produce" flow or to "absorb" it.

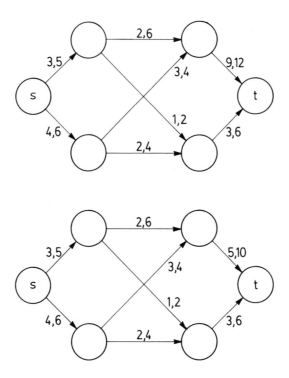

5.8 Prove that a network with lower and upper bounds on the flow in the edges has no legal flow if and only if there exists a set of vertices which neither includes the source, nor the sink, and is required to "produce" flow or to "absorb" it.

REFERENCES

[1] Ford, L. R., Jr. and Fulkerson, D. R., *Flows in Networks,* Princeton University Press, 1962.

[2] Edmonds, J., and Karp, R. M., "Theoretical Improvements in Algorithmic Efficiency for Network Flow Problems," J.ACM, Vol. 19, 1972, pp. 248-264.

[3] Dinic, E. A., "Algorithm for Solution of a Problem of Maximum Flow in a Network with Power Estimation," Soviet Math. Dokl., Vol. 11, 1970, pp. 1277-1280.

[4] Karzanov, A. V., "Determining the Maximal Flow in a Network by the Method of Preflows," *Soviet Math. Dokl.,* Vol. 15, 1974, pp. 434-437.

[5] Cherkassky, B., "Efficient Algorithms for the Maximum Flow Problem", Akad. Nauk USSR, CEMI, Mathematical Methods for the Solution of Economical Problems, Vol. 7, 1977, pp. 117-126.

[6] Malhotra, V. M., Pramodh Kumar, M., and Maheshwari, S. N., "An $O(|V|^3)$ Algorithm For Finding Maximum Flows in Networks," Computer Science Program, Indian Institute of Technology, Kanpur 208016, India. 1978.

Chapter 6

APPLICATIONS OF NETWORK
FLOW TECHNIQUES

6.1 ZERO-ONE NETWORK FLOW

Several combinatorial problems can be solved through the network flow techniques. In the networks we get, the capacity of all the edges is one. In order to get better algorithms with lower time complexities we need to study these network flow problems. We follow here the work of Even and Tarjan [1].

Consider a maximum flow problem, where for every edge e of $G(V, E)$, $c(e) = 1$.

The first observation is that in Dinic's algorithm for maximal flow in a layered network, each time we find a path, all the edges on it become blocked, and in case the last edge leads to a deadend, we backtrack on this edge and it becomes blocked. Thus, the total number of edge traversals is bounded by $|E|$ and the whole phase is of time complexity $O(|E|)$. Since the number of phases is bounded by $|V|$, the Dinic algorithm for maximum flow is of complexity $O(|V| \cdot |E|)$.

Our first goal is to prove a better bound yet: $O(|E|^{3/2})$. However, we need to prepare a few results beforehand.

Let $G(V, E)$ be a 0-1 network in which $c(e) = 1$ for all $e \in E$) with some integral legal flow function f. Define $\tilde{G}(V, \tilde{E})$ as follows:

(i) If $u \overset{e}{\rightarrow} v$ in G and $f(e) = 0$ then $e \in \tilde{E}$.
(ii) If $u \overset{e}{\leftarrow} v$ in G and $f(e) = 1$ then $u \overset{e'}{\rightarrow} v$ is in $\tilde{G}$. Clearly, e' is a new edge which corresponds to e.

Thus, $|E| = |\tilde{E}|$. Clearly, the useful edges of the layered network which is constructed for G with present flow f, with their direction of usefulness, are all edges of $\tilde{G}$.

Let us denote by $(S; \bar{S})_G$, where $s \in S$, $t \notin S$, and $\bar{S} = V - S$, the set of edges which emanate from a vertex of S and enter a vertex of $\bar{S}$, and let $c(S, G)$ be the capacity of the corresponding cut in G. Also, let M be the total maximum flow in G, while F is the total present flow.

Lemma 6.1: $\tilde{M} = M - F$.

Proof: Let S be a subset of V such that $s \in S$ and $t \notin S$. The definition of $\tilde{G}$ implies that

$$c(S, \tilde{G}) = |(S; \bar{S})_{\tilde{G}}| = \sum_{e \in (S; \bar{S})_G} (1 - f(e)) + \sum_{e \in (\bar{S}; S)_G} f(e).$$

However,

$$F = \sum_{e \in (S; \bar{S})_G} f(e) - \sum_{e \in (\bar{S}; S)_G} f(e).$$

Thus,

$$c(S, \tilde{G}) = |(S; \bar{S})_G| - F = c(S, G) - F.$$

This implies that a minimum cut of G corresponds to a minimum cut of $\tilde{G}$; i.e. is defined by the same S. By the max-flow min-cut theorem (Theorem 5.1), the capacity of a minimum cut of G is M, and the capacity of a minimum cut of $\tilde{G}$ is $\tilde{M}$ (the maximum total flow in $\tilde{G}$). Thus, the lemma follows.

Q.E.D.

Lemma 6.2: The length of the layered network for the 0-1 network defined by $G(V, E)$ (with a given s and t) and zero flow everywhere is at most $|E|/M$.

Proof: We remind the reader that V_i is the set of vertices of the ith layer of the layered network, and E_i is the set of edges from V_{i-1} to V_i. Since $f(e) = 0$ for every $e \in E$, the useful directions are all forward. Thus, every E_i is equal to $(S; \bar{S})_G$ where $S = V_0 \cup V_1 \cup \ldots \cup V_{i-1}$. Thus, by Lemma 5.1,

$$M \le |E_i| . \tag{6.1}$$

Summing up (6.1) for every $i = 1, 2, \ldots, l$ where l is the length of the layered network, we get $l \cdot M \le |E|$, or

$$l \leq |E|/M. \tag{6.2}$$

Theorem 6.1: For 0-1 networks, Dinic's algorithm is of time complexity $O(|E|^{3/2})$.

Proof: If $M \leq |E|^{1/2}$ then the number of phases is bounded by $|E|^{1/2}$, and the result follows. Otherwise, consider the phase during which the total flow reaches $M - |E|^{1/2}$. The total flow F, in $G(V, E)$, when the layered network for this phase is constructed satisfies

$$F < M - |E|^{1/2}.$$

This layered network is identical with the one constructed for $\tilde{G}$ with zero flow everywhere. Thus, by Lemma 6.1.

$$\tilde{M} = M - F > |E|^{1/2}.$$

By lemma 6.2, the length l of this layered network satisfies

$$l \leq |E|/\tilde{M} < |E|/|E|^{1/2} = |E|^{1/2}.$$

Therefore, the number of phases up to this point is at most $|E|^{1/2} - 1$, and since the number of additional phases to completion is at most $|E|^{1/2}$, the total number of phases is at most $2|E|^{1/2}$.

Q.E.D.

A 0-1 network is of *type 1* if it has no parallel edges. For such network we can prove another upper bound on the time complexity. First, we prove a lemma similar to Lemma 6.2.

Lemma 6.3: Let $G(V, E)$ define a 0-1 network of type 1 with maximum total flow M from s to t. The length l of the first layered network, when the flow is zero everywhere, is at most $2|V|/M^{1/2} + 1$.

Proof: Let V_i be the set of vertices of the ith layer. Since there are no parallel edges, the set of edges, E_{i+1}, from V_i to V_{i+1} in the layered network satisfies $|E_{i+1}| \leq |V_i| \cdot |V_{i+1}|$ for every $i = 0, 1, \ldots, l - 1$. Since each $|E_i|$ is the capacity of a cut, we get that

$$M \leq |V_i| \cdot |V_{i+1}|.$$

Thus, either $|V_i| \geq M^{1/2}$ or $|V_{i+1}| \geq M^{1/2}$. Clearly,

$$\sum_{i=0}^{l} |V_i| \leq |V|.$$

Thus,

$$\left\lfloor \frac{l+1}{2} \right\rfloor \cdot M^{1/2} \leq |\bar{V}|$$

and the lemma follows.   Q.E.D.

Theorem 6.2: For 0-1 networks of type 1, Dinic's algorithm has time complexity $O(|V|^{2/3} \cdot |E|)$.

Proof: If $M \leq |V|^{2/3}$, the result follows immediately. Let F be the total flow when the layered network, for the phase during which the total flow reaches the value $M - |V|^{2/3}$, is constructed. This layered network is identical with the first layered network for $\tilde{G}$ with zero flow everywhere. $\tilde{G}$ may not be of type 1 since it may have parallel edges, but it can have at most two parallel edges from one vertex to another; if e_1 and e_2 are antiparallel in G, $f(e_1) = 0$ and $f(e_2) = 1$, then in $\tilde{G}$ there are two parallel edges: e_1 and e_2'. A result similar to Lemma 6.3 yields that

$$l < 2^{3/2} |V| / \tilde{M}^{1/2} + 1.$$

Since $\tilde{M} = M - F > M - (M - |V|^{2/3}) = |V|^{2/3}$, we get

$$l < \frac{2^{2/3} |V|}{|V|^{1/3}} + 1 = 2^{2/3} \cdot |V|^{2/3} + 1.$$

Thus, the number of phases up to this point is $O(|V|^{2/3})$. Since the number of phases from here to completion is at most $|V|^{2/3}$, the total number of phases if $O(|V|^{2/3})$.

                                                                     Q.E.D.

In certain applications, the networks which arise satisfy the condition that for each vertex other than s or t, either there is only one edge emanating from it or only one edge entering it. Such 0-1 networks are called *type 2*.

Lemma 6.4: Let the 0-1 network defined by $G(V, E)$ be of type 2, with maximum total flow M from s to t. The length l of the first layered network, when the flow is zero everywhere, is at most $(|V| - 2)/M + 1$.

Proof: The structure of G implies that a max-flow in G can be decomposed into vertex-disjoint directed paths from s to t; i.e., no two of these paths share any vertices, except their common start-vertex s and end-vertex t. (The flow may imply some directed circuits, vertex-disjoint from each other and from the paths above, except possibly at s or t. These circuits are of no interest to us.) The number of these paths is equal to M. Let λ be the length of a shortest of these paths. Thus, each of the paths uses at least $\lambda - 1$ intermediate vertices. We have,

$$M \cdot (\lambda - 1) \leq |V| - 2,$$

which implies $\lambda \leq (|V| - 2)/M + 1$. However, $l \leq \lambda$. Thus, the lemma follows.

Q.E.D.

Lemma 6.5: If the 0-1 network defined by G is of type 2 and if the present flow function is f, then the corresponding $\tilde{G}$ defines also a type 2 0-1 network.

Proof: Clearly $\tilde{G}$ defines a 0-1 network. What remains to be shown is that in $\tilde{G}$ for every vertex v, either there is only one emanating edge or only one entering edge. If there is no flow through v (per f), then in $\tilde{G}$ v has exactly the same incident edges and the condition continues to hold. If the flow going through v is 1, (clearly, it cannot be more) assume that it enters via e_1 and leaves via e_2. In $\tilde{G}$ neither of these two edges appears, but two edges e_1' and e_2' are added, which have directions opposite to e_1 and e_2, respectively. The other edges of G which are incident to v remain intact in $\tilde{G}$. Thus, the numbers of incoming edges and outgoing edges of v remains the same. Since G is of type 2, so is $\tilde{G}$.

Q.E.D.

Theorem 6.3: For a 0-1 network of type 2, Dinic's algorithm is of time complexity $O(|V|^{1/2} \cdot |E|)$.

Proof: If $M \leq |V|^{1/2}$, then the number of phases is bounded by $|V|^{1/2}$, and the result follows. Otherwise, consider the phase during which the total flow reaches the value $M - |V|^{1/2}$. Therefore, the layered network for this phase is constructed when $F < M - |V|^{1/2}$. This layered network is identical with the first for $\tilde{G}$, with zero flow everywhere. Also, by Lemma 6.5, $\tilde{G}$ is of type 2. Thus, by Lemma 6.4, the length l of the layered network is at most $(|V| - 2)/\tilde{M} + 1$. Now, $\tilde{M} = M - F > M - (M - |V|^{1/2}) = |V|^{1/2}$.

Thus,

$$l \leq \frac{|V| - 2}{|V|^{1/2}} + 1 = O(|V|^{1/2}).$$

Therefore, the number of phases up to this one is at most $O(|V|^{1/2})$. Since the number of phases to completion is at most $|V|^{1/2}$ more, the total number of phases is at most $O(|V|^{1/2})$.

Q.E.D.

6.2 VERTEX CONNECTIVITY OF GRAPHS

Intuitively, the connectivity of a graph is the minimum number of elements whose removal from the graph disconnect it to more than one component. There are four cases. We may discuss undirected graphs or digraphs; we may discuss the elimination of edges or vertices. We shall start with the problem of determining the vertex-connectivity of an undirected graph. The other cases, which are simpler, will be discussed in the next section.

Let $G(V, E)$ be a finite undirected graph, with no self-loops and no parallel edges. A set of vertices, S, is called an (a, b) *vertex separator* if $\{a, b\} \subset V - S$ and every path connecting a and b passes through at least one vertex of S. Clearly, if a and b are connected by an edge, no (a, b) vertex separator exists. Let $a \not\rightarrow b$ mean that there is no such edge. In this case, let $N(a, b)$ be the least cardinality of an (a, b) vertex separator. Also, let $p(a, b)$ be the maximum number of pairwise vertex disjoint paths connecting a and b in G; clearly, all these paths share the two end-vertices, but no other vertex appears on more than one of them.

Theorem 6.4: If $a \not\rightarrow b$ then $N(a, b) = p(a, b)$.

This is one of the variations of Menger's theorem [2]. It is not only reminiscent of the max-cut min-flow theorem, but can be proved by it. Dantzig and Fulkerson [3] pointed out how this can be done, and we shall follow their approach.

Proof: Construct a digraph $\overline{G}(\overline{V}, \overline{E})$ as follows. For every $v \in V$ put two vertices v' and v'' in $\overline{V}$ with an edge $v' \xrightarrow{e_v} v''$. For every edge $u \xrightarrow{e} v$ in G, put two edges $u'' \xrightarrow{e'} v'$ and $v'' \xrightarrow{e''} u'$ in $\bar{G}$. Define now a network, with digraph $\bar{G}$, source a'', sink b', unit capacities for all the edges of the e_v type (let us call them *internal* edges), and infinite capacity for all the edges of the e' and e'' type (called *external* edges). For example, in Fig. 6.1(b) the network for G, as shown in Fig. 6.1(a), is demonstrated.

We now claim that $p(a, b)$ is equal to the total maximum flow F (from a'' to b') in the corresponding network. First, assume we have $p(a, b)$ vertex

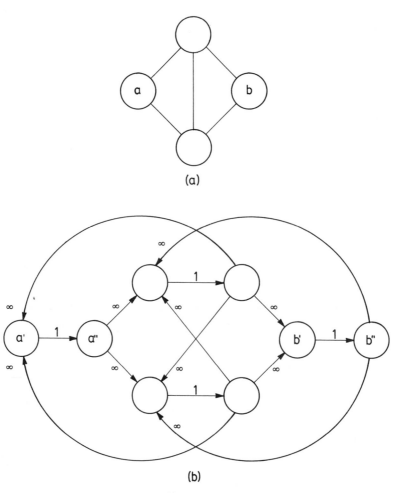

(a)

(b)

Figure 6.1

disjoint paths from a to b in G. Each such path, $a \xrightarrow{e_1} v_1 \xrightarrow{e_2} v_2 \xrightarrow{e_3} \ldots \xrightarrow{e_{l-1}} v_{l-1} \xrightarrow{e_l} b$, indicates a directed path in $\overline{G}$:

$$a'' \xrightarrow{e_1'} v_1' \xrightarrow{e_{v_1}} v_1'' \xrightarrow{e_2'} v_2' \xrightarrow{e_{v_2}} v_2'' \xrightarrow{e_3'} \ldots \xrightarrow{e_{l-1}'} v_{l-1}' \xrightarrow{e_{v_{l-1}}} v_{l-1}'' \xrightarrow{e_l'} b'.$$

These directed paths are vertex disjoint, and each can be used to flow one unit from a'' to b'. Thus,

$$F \geq p(a, b).$$

Next, assume f is a flow function which achieves a maximum total flow F in the network. We may assume that $f(e)$ is either zero or one, for every $e \in \overline{E}$. This follows from the fact that one can use the Ford and Funkerson algorithm, or the Dinic algorithm, in which the flow is alwyas integral. Also, the edges with infinite capacity enter a vertex with a single out-going edge whose capacity is one and which must satisfy the conservation rule $(C2)$, or they emanate from a vertex with only one incoming edge of unit capacity which is subject to $C2$; thus, the flow through them is actually bounded from above by one. (We have assigned them infinite capacity for convenience reasons, which will become clear shortly.) Therefore, the total flow F can be decomposed to paths, each describing the way that one unit reaches b' from a''. These paths are vertex disjoint, since the flow through v' or v'', if $v \notin \{a, b\}$, is bounded by one. Each indicates a path in G. These F paths in G are vertex disjoint too. Thus,

$$F \leq p(a, b).$$

We conclude that $F = p(a, b)$.

By the max-flow min-cut theorem, F is equal to the capacity $c(S)$ of some cut defined by some $S \subset \overline{V}$, such that $a'' \in S$ and $b' \notin S$. Since

$$c(S) = \sum_{e \in (S; \overline{S})} c(e),$$

the set $(S; \overline{S})$ consists of internal edges only. Now, every directed path from a'' to b' in $\overline{G}$ uses at least one edge of $(S; \overline{S})$. Thus, every path from a to b in G uses at least one vertex v such that $e_v \in (S; \overline{S})$. Therefore, the set $R = \{v \mid v \in V$ and $e_v \in (S; \overline{S})\}$ is an (a, b) vertex separator. Clearly $|R| = c(S)$. Thus, we have an (a, b) vertex separator whose cardinality is F. Proving that $N(a, b) \leq F = p(a, b)$.

Finally, it is easy to see that $N(a, b) \geq p(a, b)$, since every path from a to b

uses at least one vertex of the separator, and no two paths can use the same one.

Q.E.D.

The algorithm suggested in the proof, for finding $N(a, b)$, when the Dinic algorithm is used to solve the network problem, is of time complexity $O(|V|^{1/2} \cdot |E|)$. This results from the following considerations. The number of vertices in $\bar{G}$ is $2|V|$; the number of edges is $|V| + 2|E|$. Assuming $|E| \geq |V|$, we have $|\bar{V}| = O(|V|)$ and $|\bar{E}| = O(|E|)$. Since we can assign unit capacity to all the edges without changing the maximum total flow, the network is of type 2. By Theorem 6.3, the algorithm is of time complexity $O(|V|^{1/2} \cdot |E|)$. We can even find a minimum (a, b) vertex separator as follows: Once the flow is maximum, change the capacity of the external edges back to ∞ and apply the construction of the layered network. The set of vertices which appear also in this layered network, S, defines a minimum cut which consists of internal edges only. Let R be the vertices of G which correspond to the internal edges in $(S; \bar{S})$. R is a minimum (a, b) vertex separator in G. This additional work is of time complexity $O(|E|)$.

The *vertex connectivity*, c, of an undirected graph $G(V, E)$ is defined as follows:

(i) If G is completely connected, (i.e. every two vertices are connected by an edge), then $c = |V| - 1$.
(ii) If G is not completely connected then

$$c = \underset{a \neq b}{\text{Min}}\, N(a, b).$$

Lemma 6.6: If G is not completely connected then

$$\underset{a \neq b}{\text{Min}}\, p(a, b) = \underset{a, b}{\text{Min}}\, p(a, b).$$

Namely, the smallest value of $p(a, b)$ occurs also for some two vertices a and b which are not connected by an edge.

Proof: Let a, b be a pair of vertices such that $a \overset{e}{-} b$ and $p(a, b)$ is minimum over all pairs of vertices of the graph. Let G' be the graph obtained from G by dropping e. Clearly, the number of vertex disjoint paths connecting a and b in G', $p'(a, b)$, satisfies

$$p'(a, b) = p(a, b) - 1.$$

Also, since $a \not\sim b$ in G', then by Theorem 6.4, there is an (a, b) vertex separator R in G' such that $p'(a, b) = |R|$.

If $|R| = |V| - 2$, then $p(a, b) = |V| - 1$, and $p(a, b)$ cannot be the least of all $\{p(u, v)|u, v \in V\}$, since for any $u \not\sim v$, $p(u, v) \leq |V| - 2$. Hence, $|R| < |V| - 2$. Therefore, there must be some vertex $v \in V - (R \cup \{a, b\})$. Now, without loss of generality we may assume that R is also an (a, v) vertex separator (or exchange a and b). Thus, $a \not\sim v$ in G and $R \cup \{b\}$ is an (a, v) vertex separator in G. We now have

$$p(a, v) \leq |R| + 1 = p(a, b),$$

and the lemma follows.

Q.E.D.

Theorem 6.5: $c = \underset{a,b}{\text{Min}}\, p(a, b)$.

Proof: If G is completely connected then for every two vertices a and b, $p(a, b) = |V| - 1$ and the theorem holds. If G is not completely connected then, by definition

$$c = \underset{a \not\sim b}{\text{Min}}\, N(a, b).$$

By Theorem 6.4, $\underset{a \not\sim b}{\text{Min}}\, N(a, b) = \underset{a \not\sim b}{\text{Min}}\, p(a, b)$. Now by Lemma 6.6, $\underset{a \not\sim b}{\text{Min}}\, p(a, b) = \underset{a,b}{\text{Min}}\, p(a, b)$.

Q.E.D.

We can use the intermediate result,

$$c = \underset{a \not\sim b}{\text{Min}}\, p(a, b),$$

to compute the vertex connectivity of G with time complexity $O(|V|^{2.5} \cdot |E|)$. However, a slightly better bound can be obtained.

Lemma 6.7: $c \leq 2|E|/|V|$.

Proof: The vertex (or edge) connectivity of a graph cannot exceed the degree of any vertex. Thus,

$$c \leq \underset{v}{\text{Min}}\, d(v).$$

Also,

$$\sum_v d(v) = 2 \cdot |E|.$$

Thus, $\underset{v}{\text{Min}}\, d(v) \le 2 \cdot |E| / |V|$, and the lemma follows.

Q.E.D.

Let us conduct the procedure to find the vertex connectivity c, of a graph G, which is not completely connected, as follows: Order the vertices v_1, v_2, $\ldots$, $v_{|V|}$ in such a way that $v_1 \not\sim v$ for some v. Compute $N(v_1, v)$ for every v such that $v_1 \not\sim v$. (There are at most $|V|$ such computations, each of time complexity $O(|V|^{1/2} \cdot |E|)$.) Repeat the computation for v_2, v_3, $\ldots$ and terminate with v_k, once k exceeds the minimum value of $N(v_i, v)$ observed so far, γ.

Theorem 6.6: The procedure described above terminates with $\gamma = c$.

Proof: Clearly after the first computation of $N(v_1, v)$ for some $v_1 \not\sim v$, γ satisfies

$$c \le \gamma \le |V| - 2. \tag{6.3}$$

From there on γ can only decrease, but (6.3) still holds. Thus, for some $k \le |V| - 1$, the procedure will terminate. When it does, $k \ge \gamma + 1 \ge c + 1$.

By definition, c is equal to the cardinality of a minimum vertex separator R of G. Thus, at least one of the vertices v_1, v_2, $\ldots$, v_k is not in R, say v_i. R separates the remaining vertices into at least two sets, such that each path, from a vertex of one set to a vertex of another, passes through at least one vertex of R. Thus, there exists a vertex v such that $N(v_i, v) \le |R| = c$, and therefore $\gamma \le c$.

Q.E.D.

Clearly, the time complexity of this procedure is $O(c \cdot |V|^{3/2} \cdot |E|)$. By Lemma 6.7 this is bounded by $O(|V|^{1/2} \cdot |E|^2)$.

If $c = O$ then G is not connected. We can use DFS (or BFS) to test whether this is the case in $O(|E|)$ time. If $c = 1$ then G is separable, and as we have seen in Section 3.2, this can be tested also by DFS in $O(|E|)$ time. This algorithm determines also whether $c \ge 2$, i.e., whether it is non-separable. Before we discuss testing for a given k, whether $c \ge k$, let us consider the following interesting theorem about equivalent conditions for G to be nonseparable.*

Theorem 6.7: Let $G(V, E)$ be an undirected graph with $|V| > 2$ and no isolated vertices.** The following six statements are equivalent.

(1) G is nonseparable.
(2) For every two vertices x and y there exists a simple circuit which goes through both.
(3) For every two edges e_1 and e_2 there exists a simple circuit which goes through both.
(4) For every two vertices x and y and an edge e there exists a simple path from x to y which goes through e.
(5) For every three vertices x, y and z there exists a simple path from x to z which goes through y.
(6) For every three vertices x, y and z there exists a simple path from x to z which avoids y.

Proof: First we prove that (1) is equivalent to (2).

(1) $\Rightarrow$ (2): Since G is nonseparable, $c \geq 2$. By Theorem 6.5, for every two vertices x and y $p(x, y) \geq 2$. Thus, there is a simple circuit which goes through x and y.

(2) $\Rightarrow$ (1): There cannot exist a separation vertex in G since every two vertices lie on some common simple circuit.

Next, let us show that (1) and (3) are equivalent.

(1) $\Rightarrow$ (3): From G construct G' as follows. Remove the edges $u_1 \overset{e_1}{-} v_1$ and $u_2 \overset{e_2}{-} v_2$ (without removing any vertices). Add two new vertices, x and y, and four new edges: $u_1 - x - v_1$, $u_2 - y - v_2$. Clearly, none of the old vertices become separation vertices, by this change. Also, x cannot be a separation vertex, or either u_1 or v_1 are separation vertices in G. (Here $|V| > 2$ is used.) Thus, G' is nonseparable. Hence, by the equivalence of (1) and (2), G' satisfies (2). Therefore, there exists a simple circuit in G' which goes through x and y. This circuit indicates a circuit through e_1 and e_2 in G.

(3) $\Rightarrow$ (1): Let x and y be any two vertices. Since G has no isolated vertices, there is an edge e_1 incident to x and an edge e_2 incident to y. (If $e_1 = e_2$, chose any other edge to replace e_2; the replacement need not even be incident to y; the replacement exists since there is at least one other vertex, and it is not

---

*Many authors use the term *biconnected* to mean nonseparable. I prefer to call a graph biconnected if $c = 2$.
**Namely, for every $v \in V$, $d(v) > 0$. G has been assumed to have no self-loops.

isolated.) By (3) there is a simple circuit through e_1 and e_2, and therefore a circuit through x and y. Thus, (2) holds, and (1) follows:

Now, let us prove that $(3) \Rightarrow (4) \Rightarrow (5) \Rightarrow (6) \Rightarrow (3)$.

$(3) \Rightarrow (4)$: Since (3) holds, the graph G is nonseparable. Add a new edge $x \overset{e'}{-} y$, if such does not exist already in G. Clearly, the new graph, G', is still nonseparable. By the equivalence of (1) and (3), G' satisfy statement (3). Thus, there is a simple circuit which goes through e and e' in G'. Therefore, there is a simple path in G, from x to y through e.

$(4) \Rightarrow (5)$: Let e be an edge incident to vertex y; such an edge exists since there are no isolated vertices in G. By (4) there is a simple path from x to z through e. Thus, this path goes through y.

$(5) \Rightarrow (6)$: Let p be a simple path which goes from x to y through z; such a path exists since (5) holds for every three vertices. The first part of p, from x to z, does not pass through y.

$(6) \Rightarrow (1)$: If (6) holds then there cannot be any separation vertex in G.

Q.E.D.

Let us now return to the problem of testing the k connectivity of a given graph G; that is, testing whether c is greater than or equal to the given positive integer k. We have already seen that for $k = 1$ and 2, there is an $O(|E|)$ algorithm. Hopcroft and Tarjan [4] showed that $k = 3$ can also be tested in linear time, but their algorithm is quite complicated and does not seem to generalize for higher values of k. Let us present a method which was suggested by Kleitman [5] and improved by Even [6].

Let $L = \{v_1, v_2, \ldots, v_l\}$ be a subset of V, where $l \geq k$. Define $\tilde{G}$ as follows. $\tilde{G}$ includes all the vertices and edges of G. In addition it includes a new vertex s connected by an edge to each of the vertices of L. $\tilde{G}$ is called the auxiliary graph.

Lemma 6.8: Let $u \in V - L$. If in G $p(v_i, u) \geq k$ for every $v_i \in L$, then in $\tilde{G}$ $p(s, u) \geq k$.

Proof: Assume not. Then $p(s, u) < k$. By Theorem 6.4, there exists a (s, u) vertex separator S, in $\tilde{G}$, such that $|S| < k$. Let R be the set of vertices such that all paths, in $\tilde{G}$, from s to $v \in R$ pass through at least one vertex of S. Clearly, $v_i \notin R$, since v_i is connected by an edge to s. However, since $l \geq k > |S|$, there exists some $1 \leq i \leq l$ such that $v_i \notin S$. All paths from v_i to u go through vertices of S. Thus, $p(v_i, u) \leq |S| < k$, contradicting the assumption.

Q.E.D.

Let $V = \{v_1, v_2, \ldots, v_n\}$. Let j be the least integer such that for some $i < j, p(v_i, v_j) < k$ in G.

Lemma 6.9: Let j be as defined above and $\tilde{G}$ be the auxiliary graph for $L = \{v_1, v_2, \ldots, v_{j-1}\}$. In $\tilde{G}, p(s, v_j) < k$.

Proof: Consider a minimum (v_i, v_j) vertex separator S. By Theorem 6.4, $|S| < k$. Let R be the set of all vertices $v \in V$ such that all the paths from v_i to v, in G, pass through vertices of S. Clearly, $v_j \in R$. If for some $j' < j, v_{j'} \in R$, then $p(i, j') \le |S| < k$, and the choice of j is wrong. Thus, v_j is the least vertex in R (i.e. the vertex for which the subscript is minimum). Hence $L \cap R = \emptyset$. Thus, in $\tilde{G}, S$ is an (s, v_j) vertex separator, and $p(s, v_j) < k$.

$$\text{Q.E.D.}$$

The following algorithm determines whether the vertex connectivity of a given undirected graph $G(V, E)$, where $V = \{v_1, v_2, \ldots, v_n\}$, is at least k.

(1) For every i and j such that $1 \le i < j \le k$, check whether $p(v_i, v_j) \ge k$. If for some i and j this test fails then halt; G's connectivity is less then k.
(2) For every $j, k + 1 \le j \le n$ form $\tilde{G}$ (with $L = \{v_1, v_2, \ldots, v_{j-1}\}$) and check whether in $\tilde{G}$ $p(s, v_j) \ge k$. If for some j this test fails then halt; G's connectivity is less than k.
(3) Halt; the connectivity of G is at least k.

The proof of the algorithm's validity is as follows: If G's connectivity is at least k, then clearly no failure will be detected in Step (1). Also, by Lemma 6.8, no failure will occur in Step (2), and the algorithm will halt in Step (3) with the right conclusion. If G's connectivity is less than k, and it is not detected directly in Step (1) then, by Lemma 6.9, it will be detected in Step (2).

Step (1) takes $O(k^3 \cdot |E|)$ steps, since we have to solve $k(k - 1)/2$ flow problems. In each we have to find k augmenting paths, and each path takes $O(|E|)$ steps to find.

Step (2) takes $O(k \cdot |V| \cdot |E|)$ steps, since we have to solve $|V| - k$ flow problem, again for each up to total flow k.

Thus, if $k \le |V|^{1/2}$ then the time complexity of the algorithm is $O(k \cdot |V| \cdot |E|)$.

The readers who are familiar with the interesting result of Gomory and Hu [7] for finding all $|V| \cdot (|V| - 1)$ total flows, for all source-sink pairs in an undirected network by solving only $|V| - 1$ network flow problem, should notice that this technique is of no help in our problem. The reason for that is that even if G is undirected, the network we get for vertex connectivity testing is directed.

6.3 CONNECTIVITY OF DIGRAPHS AND EDGE CONNECTIVITY

First, let us consider the problem of vertex-connectivity of a digraph $G(V, E)$. The definition of an (a, b) vertex separator is the same as in the undirected case, except that now all we are looking at are directed paths from a to b; i.e., an (a, b) *vertex separator* is a set of vertices S such that $\{a, b\} \cap S = \emptyset$ and every directed path from a to b passes through at least one vertex of S. Accordingly, $N(a, b)$ and $p(a, b)$ are defined. The theorem analogous to Theorem 6.4, still holds, except that the algorithm is even simpler: For every edge $u \xrightarrow{e} v$ in G there is only one edge $u'' \xrightarrow{e'} v'$ in $\overline{G}$.

The *vertex connectivity, c,* of a digraph $G(V, E)$ is defined as follows:

(i) If G is completely connected, (i.e. for every two vertices a and b there are edges $a \rightarrow b$ and $b \rightarrow a$), then $c = |V| - 1$.

(ii) If G is not completely connected then

$$c = \operatorname*{Min}_{a \not\vdash b} N(a, b).$$

The lemma analogous to Lemma 6.6 still holds, and the proof goes along the same lines. Also, the theorem analogous to Theorem 6.5 holds, and the complexity it yields is the same. If G has no parallel edges a statement like Lemma 6.7 holds and the procedure and the proof of its validity (Theorem 6.6) extend to the directed case, except that for each v_i we compute both $N(v_i, v)$ and $N(v, v_i)$.

The algorithm for testing k connectivity extends also to the directed case and again all we need to change is that whenever $p(a, b)$ was computed, we now have to compute both $p(a, b)$ and $p(b, a)$.

Let us now consider the case of edge connectivity both in graphs and digraphs.

Let $G(V, E)$ be an undirected graph. A set of edges, T, is called an (a, b) *edge separator* if every path from a to b passes through at least one edge of T. Let $M(a, b)$ be the least cardinality of an (a, b) edge separator. Let $p(a, b)$ be now the maximum number of edge disjoint paths which connect a with b.

Theorem 6.8: $M(a, b) = p(a, b)$.

The proof is similar to that of Theorem 6.4, only simpler. There is no need to split vertices. Thus, in $\bar{G}$, $\bar{V} = V$. We still represent each edge $u - v$ of G by two edges $u \xrightarrow{e'} v$ and $v \xrightarrow{e''} u$ in $\bar{G}$. There is no loss of generality in assuming that the flow function in $\bar{G}$ satisfies the condition that either $f(e') = 0$ or $f(e'') = 0$; for if $f(e') = f(e'') = 1$ then replacing both by 0 does not change the total flow. The rest of the proof raises no difficulties.

The *edge connectivity, c,* of a graph G is defined by $c = \text{Min}_{a,b} M(a, b)$. By Theorem 6.8 and its proof, we can find c by the network flow technique. The networks we get are of type 1. Both Theorem 6.1 and Theorem 6.2 apply. Thus, each network flow problem is solvable by Dinic's algorithm with complexity $O(\text{Min} \{|E|^{3/2}, |V|^{2/3} \cdot |E|\})$.

Let T be a minimum edge separator in G; i.e. $|T| = c$. Let v be any vertex of G. For every vertex v', on the other side of T, $M(v, v') = c$. Thus, in order to determine c, we can use

$$c = \underset{v' \in V - \{v\}}{\text{Min}} M(v, v').$$

We need to solve at most $|V| - 1$ network flow problems. Thus, the complexity of the algorithm is $O(|V| \cdot |E| \cdot \text{Min} \{|E|^{1/2}, |V|^{2/3}\})$.

In the case of edge connectivity of digraphs, we need to consider directed paths. The definition of an (a, b) *edge separator* is accordingly a set of edges, T, such that every directed path from a to b uses at least one edge of T. The definition of $p(a, b)$ again uses directed paths, and the proof of the statement analogous to Theorem 6.8 is the easiest of all, since $\bar{G}$ is now G with no changes.

In the definition of c, the edge connectivity, we need the following change:

$$c = \text{Min} \{M(a, b)|(a, b) \in V \times V\},$$

namely, we need to consider all ordered pairs of vertices.

The networks we get are still of type 1 and the complexity of each is still $O(|E| \cdot \text{Min}\{|E|^{1/2}, |V|^{2/3}\})$. The approach of testing for one vertex v, both $M(v, v')$ and $M(v', v)$ for all $v' \in V - \{v\}$ still works, to yield the same complexity: $O(|V| \cdot |E| \cdot \text{Min}\{|E|^{1/2}, |V|^{2/3}\})$. However, the same result, with an improvement only in the constant coefficient follows from the following interesting observation of Schnorr [8], which applies both to the directed and undirected edge connectivity problems.

Lemma 6.10: Let $v_1, v_2, \ldots, v_n$ be a circular ordering (i.e. $v_{n+1} = v_1$) of the vertices of a digraph G. The edge connectivity, c, of G satisfies

$$c = \underset{1 \leq i \leq n}{\text{Min}} M(v_i, v_{i+1}).$$

Proof: Let T be a minimum edge separator in G. That means that there are two vertices a and b such that T is an (a, b) edge separator. Define

$L = \{v|$ there is a directed path from a to v which avoids $T\}$,

$R = \{v|$ there is no directed path from a to v which avoids $T\}$.

Clearly, $L \cup R = V$ and $L \cap R = \emptyset$. Let $l \in L$ and $r \in R$. T is an (l, r) edge separator; for if it is not, then r belongs in L. Therefore, $M(l, r) \leq |T|$. Since T is a minimum edge separator, $M(l, r) = |T|$. Now neither L nor R are empty, since they contain a and b, respectively.

Consider now the circular ordering of V. There must be an i, $1 \leq i \leq n$, such that $v_i \in L$ and $v_{i+1} \in R$. Hence, the result.

Q.E.D.

Both, in the case of graphs and digraphs we can test for k connectivity, easily, in time complexity $O(k \cdot |V| \cdot |E|)$. Instead of running each network flow problem to completion, we terminate it when the total flow reaches k. Each augmenting path takes $O(|E|)$ time and there are $|V|$ flow problems. As we can see, testing for k edge connectivity is much easier than for k vertex connectivity. The reason is that vertices cannot participate in the separating set which consists of edges.

We can also use this approach to determine the edge connectivity, c, in time $O(c \cdot |V| \cdot |E|)$. We run all the $|V|$ network flow problems in parallel, one augmenting path for each network in turn. When no augmenting path exists in any of the $|V|$ problems, we terminate. The cost increase is only in space, since we need to store all $|V|$ problems simultaneously. One can use binary search on c to avoid this increase in space requirements, but in this case the time complexity is $O(c \cdot |V| \cdot |E| \cdot \log c)$.

We conclude our discussion of edge connectivity with a very powerful theorem of Edmonds [9]. The proof presented here is due to Lovász [10].

Theorem 6.9: Let a be a vertex of a digraph $G(V, E)$ and $\text{Min}_{v \in V - \{a\}} M(a, v)$ $= k$. There are k edge disjoint directed spanning trees of G rooted at a.

Proof: The theorem trivially holds for $k = 1$. We prove the theorem by induction on k. Let us denote by $\delta_G(S)$ the number of edges in $(S; \bar{S})$ in G. If H

is a subgraph of G then $G - H$ is the digraph resulting from the deletion of all the edges of H from G.

Clearly, the condition that $\text{Min}_{v \in V - \{a\}} M(a, v) \geq k$ is equivalent to the statement that for every $S \subset V$, $S \neq V$ and $a \in S$, $\delta_G(S) \geq k$.

We prove the theorem by induction on k.

Let $F(V', E')$ be a subgraph of G such that
 (i) F is a directed tree rooted at a (which is not necessarily spanning).
(ii) For every $S \subset V$, $S \neq V$ and $a \in S$, $\delta_{G-F}(S) \geq k - 1$.

If F is a spanning directed tree then we get the result immediately; by the inductive hypothesis there are $k - 1$ edge disjoint spanning trees rooted at a in $G - F$, and F is one more.

The crux of the proof is to show that if F is not spanning then an edge of the set $(V'; \overline{V}')$ can be added to F, to increase its number of vertices by one and still satisfy both (i) and (ii).

Consider the following three conditions on a subset of vertices, S:

(1) $a \in S$,
(2) $S \cup V' \neq V$,
(3) $\delta_{G-F}(S) = k - 1$

Let us show that if no such S exists then one can add any edge $e \in (V'; \overline{V}')$ to F. Clearly, $F + e$ satisfies (i). Now, if (ii) does not hold then there exists an S such that $S \neq V$, $a \in S$ and $\delta_{G-(F+e)}(S) < k - 1$. It follows that $\delta_{G-F}(S) < k$. Now, by (ii), $\delta_{G-F}(S) \geq k - 1$. Thus, $\delta_{G-F}(S) = k - 1$, and S satisfies condition (3). Let u and v be vertices such that $u \overset{e}{\rightarrow} v$. Since $\delta_{G-(F+e)}(S) < k - 1$ and $\delta_{G-F}(S) = k - 1$, $v \notin S$. Also, $v \notin V'$. Thus $S \cup V' \neq V$, satisfying condition (2). Therefore, S satisfies all three conditions; a contradiction.

Now, let A be a maximal* set of vertices which satisfies (1), (2) and (3). Since the edges of F all enter vertices of V'

$$\delta_{G-F}(A \cup V') = \delta_G(A \cup V') \geq k.$$

By condition (3)

$$\delta_{G-F}(A \cup V') > \delta_{G-F}(A).$$

---

*Namely, no larger set which contains A has all three properties.

The inequality implies that there exists an edge $x \overset{e}{\to} y$ which belongs to $(A \cup V'; \overline{A \cup V'})$ and does not belong to $(A; \overline{A})$ in $G - F$. Hence, $x \in A \cap V'$ and $y \in \overline{A} \cap \overline{V}'$. Clearly $F + e$ satisfies (i). It remains to be shown that it satisfies (ii).

Let $S \subset V$, $S \neq V$ and $a \in S$. If $e \notin (S; \overline{S})$ then

$$\delta_{G-(F+e)}(S) = \delta_{G-F}(S) \geq k - 1.$$

Assume $e \in (S; \overline{S})$. It is not hard to prove that for every two subsets of V, S and A

$$\delta_{G-F}(S \cup A) + \delta_{G-F}(S \cap A) \leq \delta_{G-F}(S) + \delta_{G-F}(A),$$

by considering the sets of edges connecting $S \cap A$, $S \cap \overline{A}$, $\overline{S} \cap A$ and $\overline{S} \cap \overline{A}$. Now, $\delta_{G-F}(A) = k - 1$ and $\delta_{G-F}(S \cap A) \geq k - 1$. Therefore,

$$\delta_{G-F}(S \cup A) \leq \delta_{G-F}(S).$$

Since $x \in S$ and $x \notin A$, $S \nsubseteq A$. Namely $S \cup A$ is larger than A. Also, $y \in \overline{S}$, $y \in \overline{A}$ and $y \in \overline{V}'$. Thus, $(S \cup A) \cup V' \neq V$. By the maximality of A, $\delta_{G-F}(S \cup A) \geq k$. This implies that $\delta_{G-F}(S) \geq k$ and therefore $\delta_{G-(F+e)}(S) \geq k - 1$, proving (ii).

$$\text{Q.E.D.}$$

The proof provides an algorithm for finding k edge disjoint directed trees rooted at a. We look for a tree F such that $\text{Min}_{v \in V - \{a\}} M(a, v) \geq k - 1$ in $G - F$, by adding to F one edge at a time. For each candidate edge e we have to check whether $\text{Min}_{v \in V - \{a\}} M(a, v) \geq k - 1$ in $G - (F + e)$. This can be done by solving $|V| - 1$ network flow problems, each of complexity $O(k \cdot |E|)$. Thus, the test for each candidate edge is $O(k \cdot |V| \cdot |E|)$. No edge need be considered more than once in the construction of F, yielding the time complexity $O(k \cdot |V| \cdot |E|^2)$. Since we repeat the construction k times, the whole algorithm is of time complexity $O(k^2 \cdot |V| \cdot |E|^2)$.

The following theorem was conjectured by Y. Shiloach and proved by Even, Garey and Tarjan [11].

Theorem 6.10: Let $G(V, E)$ be a digraph whose edge connectivity is at least k. For every two vertices, u and v, and every l, $0 \leq l \leq k$, there are l directed paths from u to v and $k - l$ directed paths from v to u which are all edge disjoint.

Proof: Construct an auxiliary graph $\bar{G}$ by adding to G a new vertex a, l parallel edges from a to u and $k - l$ parallel edges from a to v. Let us first show that, in $\bar{G}$ $\text{Min}_{\omega \in V} M(a, \omega) = k$.

If $\text{Min}_{\omega \in V} M(a, \omega) < k$ then there exists a set S such that $S \subset V \cup \{a\}$, $a \in S$ and $|(S; \bar{S})| < k$ in $\bar{G}$. Clearly $S \neq \{a\}$ for $|(\{a\}; V)| = k$. Let $x \in S - \{a\}$. Thus, $M(x, y) < k$ for every $y \in \bar{S}$, and the same holds in G too, since G is a subgraph of $\bar{G}$. This contradicts the assumption that in G the edge connectivity is at least k.

Now, by Theorem 6.9, there are k edge disjoint directed spanning trees, in $\bar{G}$, rooted at a. Exactly one edge out of a appears in each tree. Thus, each of the trees which uses an edge $a \rightarrow u$ contains a directed path from u to v, and each of the trees which uses an edge $a \rightarrow v$ contains a directed path from v to u. All these paths are, clearly, edge disjoint.

Q.E.D.

Corollary 6.1: If the edge connectivity of a digraph is at least 2 then for every two vertices u and v there exists a directed circuit which goes through u and v in which no edge appears more than once.

Proof: Use $k = 2$, $l = 1$ in Theorem 6.10.

Q.E.D.

It is interesting to note that no such easy result exists in the case of vertex connectivity and simple directed circuit through given two vertices. In Reference [11], a digraph with vertex connectivity 5 is shown such that for two of its vertices there is no simple directed circuit which passes through both. The author does not know whether any vertex connectivity will guarantee the existence of a simple directed circuit through any two vertices.

6.4 MAXIMUM MATCHING IN BIPARTITE GRAPHS

A set of edges, M, of a graph $G(V, E)$ with no self-loops, is called a *matching* if every vertex is incident to at most one edge of M. The problem of finding a maximum matching was first solved in polynomial time by Edmonds [12]. The best known result of Even and Kariv [13] is $O(|V|^{2.5})$. These algorithms are too complicated to be included here, and they do not use network flow techniques.

An easier problem is to find a maximum matching in a *bipartite* graph, i.e., a graph in which $V = X \cup Y$, $X \cap Y = \emptyset$ and each edge has one end vertex in X and one in Y. This problem is also known as the marriage problem. We shall present here its solution via network flow and show that its complexity is $O(|V|^{1/2} \cdot |E|)$. This result was first achieved by Hopcroft and Karp [14].

Let us construct a network $N(G)$. Its digraph $\overline{G}(\overline{V}, \overline{E})$ is defined as follows:

$$\overline{V} = \{s, t\} \cup V,$$

$$\overline{E} = \{s \rightarrow x | x \in X\} \cup \{y \rightarrow t | y \in Y\} \cup \{x \rightarrow y | x - y \text{ in } G\}.$$

Let $c(s \rightarrow x) = c(y \rightarrow t) = 1$ for every $x \in X$ and $y \in Y$. For every edge $x \overset{e}{\rightarrow} y$ let $c(e) = \infty$. (This infinite capacity is defined in order to simplify our proof of Theorem 6.12. Actually, since there is only one edge entering x, with unit capacity, the flow in $x \rightarrow y$ is bounded by 1.) The source is s and the sink is t. For example consider the bipartite graph G shown in Fig. 6.2(a). Its corresponding network is shown in Fig. 6.2(b).

Theorem 6.11: The number of edges in a maximum matching of a bipartite graph G is equal to the maximum flow, F, in its corresponding network, $N(G)$.

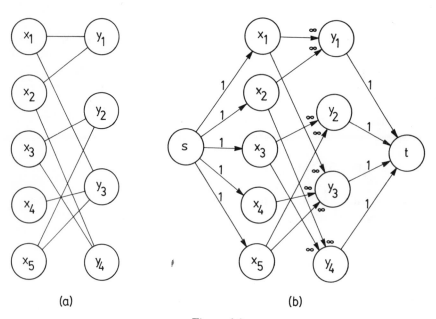

(a)   (b)

Figure 6.2

Proof: Let M be a maximum matching. For each edge $x \to y$ of M, use the directed path $s \to x \to y \to t$ to flow one unit from s to t. Clearly, all these paths are vertex disjoint. Thus, $F \geq |M|$.

Let f be a flow function of $N(G)$ which is integral. (There is no loss of generality here, since we saw, in Chapter 5, that every network with integral capacities has a maximum integral flow.) All the directed path connecting s and t are of the form $s \to x \to y \to t$. If such a path is used to flow (one unit) from s to t then no other edge $x \to y'$ or $x' \to y$ can carry flow, since there is only one edge $s \to x$ and its capacity is one, and the same is true for $y \to t$. Thus, the set of edges $x \to y$, for which $f(x \to y) = 1$, indicates a matching in G. Thus, $|M| \geq F$.

Q.E.D.

The proof indicates how the network flow solution can yield a maximum matching. For our example, a maximum flow, found by Dinic's algorithm is shown in Fig. 6.3(a) and its corresponding matching is shown in Fig. 6.3(b).

The algorithm, of the proof, is $O(|V|^{1/2} \cdot |E|)$, by Theorem 6.3, since the network is, clearly, of type 2.

Next, let us show that one can also use the max-flow min-cut theorem to prove a theorem of Hall [15]. For every $A \subset X$, let $\Gamma(A)$ denote the set of ver-

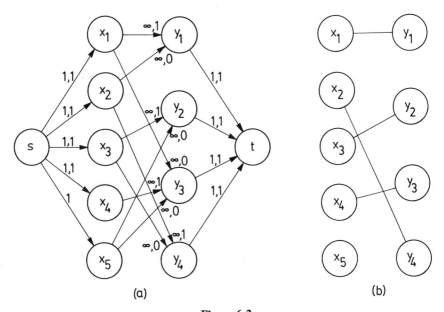

(a)   (b)

Figure 6.3

tices (all in Y) which are connected by an edge to a vertex of A. A matching, M, is called *complete* if $|M| = |X|$.

Theorem 6.12: A bipartite graph G has a complete matching if and only if for every $A \subset X$, $|\Gamma(A)| \geq |A|$.

Proof: Clearly, if G has a complete matching M, then each x has a unique "mate" in Y. Thus, for every $A \subset X$, $|\Gamma(A)| \geq |A|$.

Assume now that G does not have a complete matching. Let S be the set of labeled vertices (in the Ford and Fulkerson algorithm, or Dinic's algorithm) upon termination. Clearly, the maximum total flow is equal to $|M|$, but $|M| < |X|$. Let $A = X \cap S$. Since all the edges of the type $x \rightarrow y$ are of infinite capacity, $\Gamma(A) \subset S$. Also, no vertex of $Y - \Gamma(A)$ is labeled, since there is no edge connecting it to a labeled vertex. We have

$$(S; \bar{S}) = (\{s\}; X - A) \cup (\Gamma(A); \{t\}).$$

Since $|(S; \bar{S})| = |M| < |X|$, we get

$$|X - A| + |\Gamma(A)| < |X|,$$

which implies $|\Gamma(A)| < |A|$.

Q.E.D.

6.5   TWO PROBLEMS ON PERT* DIGRAPHS

A *PERT digraph* is a finite digraph $G(V, E)$ with the following properties:

(i) There is a vertex s, called the *start vertex,* and a vertex $t(\neq s)$, called the *termination vertex.*
(ii) G has no directed circuits.
(iii) Every vertex $v \in V - \{s, t\}$ is on some directed path from s to t.

A PERT digraph has the following interpretation. Every edge represents a process. All the processes which are represented by edges of $\beta(s)$, can be started right away. For every vertex v, the processes represented by edges of $\beta(v)$ can be started when all the processes represented by edges of $\alpha(v)$ are completed.

----

*Program Evaluation and Review Technique.

Our first problem deals with the question of how soon can the whole project be completed; i.e., what is the shortest time, from the moment the processes represented by $\beta(s)$ are started, until all the processes represented by $\alpha(t)$ are completed. We assume that the resources for running the processes are unlimited. For this problem to be well defined let us assume that each $e \in E$ has an assigned *length* $l(e)$, which specifies the time it takes to execute the process represented by e. The minimum completion time can be found by the following algorithm:

(1) Assign s the lable 0 ($\lambda(s) \leftarrow 0$). All other vertices are "unlabeled".
(2) Find a vertex, v, such that v is unlabeled and all edges of $\alpha(v)$ emanate from labeled vertices. Assign

$$\lambda(v) \leftarrow \underset{\substack{e \\ u \rightarrow v}}{\text{Max}} \{\lambda(u) + l(e)\}. \tag{6.4}$$

(3) If $v = t$, halt; $\lambda(t)$ is the minimum completion time. Otherwise, go to Step (2).

In Step (2), the existence of a vertex v, such that all the edges of $\alpha(v)$ emanate from labeled vertices is guaranteed by Condition (ii) and (iii): If no unlabeled vertex satisfies the condition then for every unlabeled vertex, v, there is an incoming edge which emanates from another unlabeled vertex. By repeatedly tracing back these edges, one finds a directed circuit. Thus, if no such vertex is found then we conclude that either (ii) or (iii) does not hold.

It is easy to prove, by induction on the order of labeling, that $\lambda(v)$ is the minimum time in which all processes, represented by the edges of $\alpha(v)$, can be completed.

The time complexity of the algorithm can be kept down to $O(|E|)$ as follows: For each vertex, v, we keep count of its incoming edges from unlabeled vertices; this count is initially set to $d_{in}(v)$; each time a vertex, u, gets labeled we use the list $\beta(u)$ to decrease the count for all v such that $u \rightarrow v$, accordingly; once the count of a vertex v reaches 0, it enters a queue of vertices to be labeled.

Once the algorithm terminates, by going back from t to s, via the edge which determined the label of the vertex, we can trace a longest path from s to t. Such a path is called *critical*.* Clearly, there may be more than one critical path. If one wants to shorten the completion time, $\lambda(t)$, then on each critical path at least one edge length must be shortened.

---

*The whole process is sometimes called the Critical Path Method (CPM).

Next, we shall consider another problem concerning PERT digraphs, in which there is no reference to edge lengths. Assume that each of the processes, represented by the edges, uses one processor for its execution. The question is: How many processors do we need in order to be sure that no execution will ever be delayed because of shortage of processors? We want to avoid such a delay without relying on the values of $l(e)$'s either because they are unknown or because they vary from time to time.

Let us solve a minimum flow problem in the network whose digraph is G, source s, sink t, lower bound $b(e) = 1$ for all $e \in E$ and no upper bound (i.e. $c(e) = \infty$ for all $e \in E$). Condition (iii) assures the existence of a legal flow (see Problem 5).

For example, consider the PERT digraph of Fig. 6.4. The minimum flow (which in this case is unique) is shown in Figure 6.5(a), where a maximum cut is shown too.

A set of edges is called *concurrent* if for no two edges in the set there is a directed path which passes through both. Now, let T be the set of vertices which are labeled in the last attempt to find an augmenting path from t to s. Clearly, $t \in T$ and $s \notin T$. The set of edges $(\overline{T}; T)$ is a maximum cut; there are no edges in $(T; \overline{T})$, for there is no upper bound on the flow in the edges, and any such edge would enable to continue the labeling of vertices. Thus, the set $(\overline{T}; T)$ is concurrent.

If S is a set of concurrent edges then the number of processors required is at least $|S|$. This can be seen by assigning the edges of S a very large length,

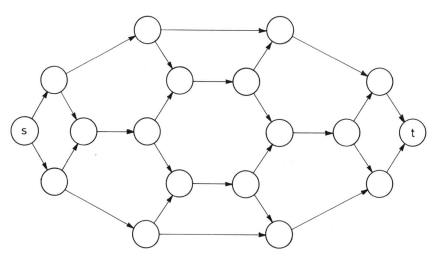

Figure 6.4

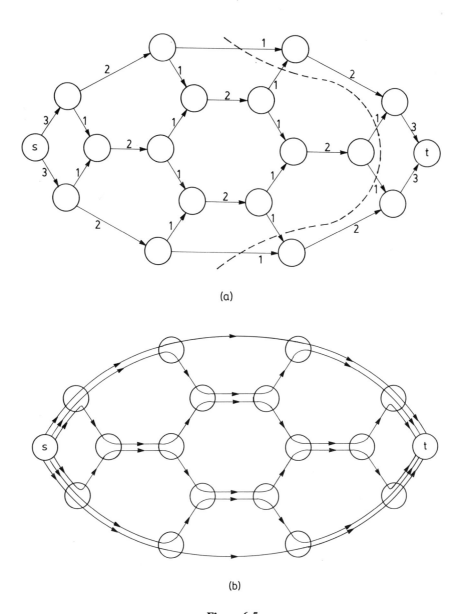

(a)

(b)

Figure 6.5

and all the others a short length. Since no directed path leads from one edge of S to another, they all will be operative simultaneously. This implies that the number of processors required is at least $|(\overline{T}; T)|$.

However, the flow can be decomposed into F directed paths from s to t, where F is the minimum total flow, such that every edge is on at least one such path (since $f(e) \geq 1$ for every $e \in E$). This is demonstrated for our example in Fig. 6.5(b). We can, now, assign to each processor all the edges of one such path. Each such processor executes the processes, represented by the edges of the path in the order in which they appear on the path. If one process is assigned to more than one processor, then one of them executes while the others are idle. It follows that whenever a process which corresponds to $u \rightarrow v$, is executable (since all the the processes which correspond to $\alpha(u)$ have been executed), the processor to which this process is assigned is available for its execution. Thus, F processors are sufficient for our purpose.

Since $F = |(\overline{T}; T)|$, by the min-flow max-cut theorem, the number of processors thus assigned is minimum.

The complexity of this procedure is as follows. We can find a legal initial flow in time $O(|V| \cdot |E|)$, by tracing for each edge a directed path from s to t via this edge, and flow through it one unit. This path is found by starting form the edge, and going forward and backward from it until s and t are reached. Next, we solve a maximum flow problem, from t to s, by the algorithm of Dinic, using MPM, in time $O(|V|^3)$. Thus, the whole procedure is of complexity $O(|V|^3)$, if $|E| \leq |V|^2$.

PROBLEMS

6.1 Let $G(V, E)$ be an acyclic* finite digraph. We wish to find a minimum number of directed vertex-disjoint paths which cover all the vertices; i.e., every vertex is on exactly one path. The paths may start anywhere and end anywhere, and their lengths are not restricted in any way. A path may be of zero length; i.e. consist of one vertex.

(a) Describe an algorithm for achieving this goal, and make it as efficient as possible. (Hint. Form a network as follows:

---

*A digraph is called acyclic if it has no directed circuits.

$$V' = \{s, t\} \cup \{x_1, x_2, \ldots, x_{|V|}\} \cup \{y_1, y_2, \ldots, y_{|V|}\}$$
$$E' = \{s \rightarrow x_i | 1 \leq i \leq |V|\} \cup \{y_i \rightarrow t | i \leq i \leq |V|\} \cup$$
$$\{x_i \rightarrow y_j | v_i \rightarrow v_j \text{ in } G\}.$$

The capacity of all edges is 1.

Show that the minimum number of paths which cover V in G is equal to $|V| - F$ where F is the maximum total flow of the network.)

(b) Is the condition that G is acyclic essential for the validity of your algorithm? Explain.

(c) Give the best upper bound you can on the time complexity of your algorithm.

6.2 This problem is similar to 6.1, except that the paths are not required to be vertex (or edge) disjoint.

(a) Describe an algorithm for finding a minimum number of covering paths. (Hint. Form a network as follows:

$$V' = \{s, t\} \cup \{x_1, x_2, \ldots, x_{|V|}\} \cup \{y_1, y_2, \ldots, y_{|V|}\}$$
$$E' = \{s \rightarrow x_i | 1 \leq i \leq |V|\} \cup \{y_i \rightarrow t | 1 \leq i \leq |V|\} \cup$$
$$\{x_i \rightarrow y_i | 1 \leq i \leq |V|\} \cup \{y_i \rightarrow x_j | v_i \rightarrow v_j \text{ in } G\}.$$

The lower bound of each $x_i \rightarrow y_i$ edge is 1.
The lower bound of all other edges is 0.
The upper bound of all the edges is ∞. Find a minimum flow from s to t.)

(b) Is the condition that G is acyclic essential for the validity of your algorithm? Explain.

(c) Give the best upper bound you can on the time complexity of your algorithm. (Hint. $O(|V| \cdot |E|)$ is achievable.)

(d) Two vertices u and v are called *concurrent* if no directed path exist from u to v, or from v to u. A set of concurrent vertices is such that every two in the set are concurrent. Prove that the minimum number of paths which cover the vertices of G is equal to the maximum number of concurrent vertices. (This is Dilworth's Theorem [16].)

6.3 (a) Let $G(X, Y, E)$ be a finite bipartite graph. Describe an efficient algorithm for finding a minimum set of edges such that each vertex is an end vertex of at least one of the edges in the set.

(b) Discuss the time complexity of your algorithm.

(c) Prove that the size of a minimum set of edges which cover the vertices (as in (a)) is equal to the maximum size of an independent set* of vertices of G.

6.4 (a) Prove that if $G(X, Y, E)$ is a complete bipartite graph (i.e. for every two vertices $x \in X$ and $y \in Y$, there is an edge $x - y$) then the vertex connectivity of G is

$$c(G) = \text{Min}\{|X|, |Y|\}.$$

(b) Prove that for every k there exists a graph G such that $c(G) \geq k$ and G has no Hamilton path. (See Problem 1.2.)

6.5 Let M be a matching of a bipartite graph. Prove that there exists a maximum matching M' such that every vertex which is matched in M is matched also in M'.

6.6 Let $G(V, E)$ be a finite acyclic digraph with exactly one vertex s for which $d_{in}(s) = 0$ and exactly one vertex t for which $d_{out}(t) = 0$. We say that the edge $a \rightarrow b$ is greater than the edge $c \rightarrow d$ if and only if there is a directed path, in G, from b to c. A set of edges is called a *slice* if no edge in it is greater than another, and it is maximal; no other set of edges with this property contains it. Prove that the following three conditions on a set of edges, P, are equivalent.:

(1) P is a slice.

(2) P is an (s, t) edge separator in which no edge is greater than any other.

(3) $P = (S; \bar{S})$ for some $\{s\} \subset S \subset V - \{t\}$ such that $(\bar{S}; S) = \emptyset$.

6.7 (The problem of a System of Distinct Representatives). Let $S_1, S_2, \ldots,$ S_m be finite sets. A set $\{e_1, e_2, \ldots, e_m\}$ is called an SDR if for every $1 \leq i \leq m, e_i \in S_i$.

(a) Describe an efficient algorithm for finding an SDR, if one exists. (Hint. Define a bipartite graph and solve a matching problem.)

(b) Prove that an SDR exists if and only if the union of any $1 \leq k \leq m$ of the sets contains at least k elements.

---

*A set of vertices of a graph is called *independent* if there is no edge between two vertices of the set.

6.8 Let π_1 and π_2 be two partitions of a set of m elements, each containing exactly r disjoint subsets. We want to find a set of r elements such that each of the subsets of π_1 and π_2 is represented.

    (a) Describe an efficient algorithm to determine whether there is such a set of r representatives.

    (b) State a necessary and sufficient condition for the existence of such a set, similar to Theorem 6.12.

6.9 Let $G(V, E)$ be a completely connected digraph (see Problem 1.3); it is called *classifiable* if V can be partitioned into two nonempty classes, A and B, such that all the edges connecting between them are directed from A to B. Let $V = \{v_1, v_2, \ldots, v_n\}$ where the vertices satisfy

$$d_{\text{out}}(v_1) \leq d_{\text{out}}(v_2) \leq \cdots \leq d_{\text{out}}(v_n).$$

Prove that G is classifiable if and only if there exists a $k < n$ such that

$$\sum_{i=1}^{k} d_{\text{out}}(v_i) = \binom{k}{2}.$$

6.10 Let S be a set of people such that $|S| \geq 4$. We assume that acquaintance is a mutual relationship. Prove that if in every subset of 4 people there is one who knows all the others then there is someone in S who knows everybody.

6.11 In the acyclic digraph, shown below, there are both AND vertices (de-

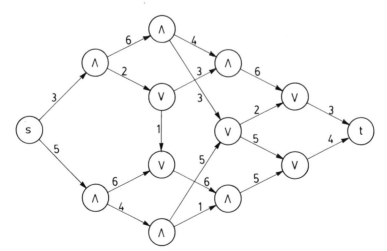

signated by ∧) and OR vertices (designated by ∨). As in a PERT network, the edges represent processes, and the edge length is the time the process requires. The processes which are represented by the edges which emanate from an AND (OR) vertex can be started when all (at least one of) the incoming processes are (is) completed. Describe an algorithm for finding the minimum time from start (s) to reach termination (which depends on the type of t). Apply your algorithm on the given network. What is the complexity of your algorithm.

6.12 Consider Problem 6.11 on digraphs which are not necessarily acyclic. Show how to modify the algorithm to solve the problem, or conclude that there is no solution.

6.13 In a school there are n boys and n girls. Each boy knows exactly k girls ($1 \leq k \leq n$) and each girl knows exactly k boys. Prove that if "knowing" is mutual then all the boys and girls can participate in one dance, where every pair of dancers (a boy and a girl) know each other. Also show that it is always true that k consecutive dances can be organized so that everyone will dance once with everyone he, or she, knows.

6.14 Prove or disprove the following claim: If the vertex connectivity of a digraph is at least 2 then for every three vertices x, y, z there exists a simple directed path from x to z via y.

6.15 Let G be a finite undirected graph whose edge connectivity is at least 2. For each one of the following claims determine whether it is true or false. Justify your answer.

    (a) For every three vertices x, y, z there exists a path, in which no edge appears more than once, from x to z via y.

    (b) For every three vertices x, y, z there exists a path, in which no edge appears more than once, from x to z which avoids y.

REFERENCES

[1] Even, S., and Tarjan, R. E., "Network Flow and Testing Graph Connectivity," SIAM J. on Computing, Vol. 4, 1975, pp. 507–518.

[2] Menger, K., "Zur Allgemeinen Kurventheorie", Fund. Math., Vol. 10, 1927, pp. 96–115.

[3] Dantzig, G. B., and Fulkerson, D. R., "On the Max-Flow Min-Cut Theorem of Networks", *Linear Inequalities and Related Systems,* Annals of Math. Study 38, Princeton University Press, 1956, pp. 215–221.

[4] Hopcroft, J., and Tarjan, R. E., "Dividing a Graph into Triconnected Components", SIAM J. on Computing, Vol. 2, 1973, pp. 135–158.

[5] Kleitman, D. J., "Methods for Investigating Connectivity of Large Graphs", IEEE Trans. on Circuit Theory, CT-16, 1969, pp. 232–233.

[6] Even, S., "Algorithm for Determining whether the Connectivity of a Graph is at Least k." SIAM J. on Computing, Vol. 4, 1977, pp. 393–396.

[7] Gomory, R. E., and Hu, T. C., "Multi-Terminal Network Flows", J. of SIAM, Vol. 9, 1961, pp. 551–570.

[8] Schnorr, C. P., "Multiterminal Network Flow and Connectivity in Unsymmetrical Networks", Dept. of Appl. Math, University of Frankfurt, Oct. 1977.

[9] Edmonds, J., "Edge-Disjoint Branchings", in *Combinatorial Algorithms, Courant Inst. Sci. Symp. 9,* R. Rustin, Ed., Algorithmics Press Inc., 1973, pp. 91–96.

[10] Lovász, L., "On Two Minimax Theorems in Graph Theory", to appear in J. of Combinatorial Th.

[11] Even, S., Garey, M. R. and Tarjan, R. E., "A Note on Connectivity and Circuits in Directed Graphs", unpublished manuscript (1977).

[12] Edmonds, J., "Paths, Trees, and Flowers", Canadian J. of Math., Vol. 17, 1965, pp. 449–467.

[13] Even, S. and Kariv, O., "An $O(n^{2.5})$ Algorithm for Maximum Matching in General Graphs", *16-th Annual Symp. on Foundations of Computer Science,* IEEE, 1975, pp. 100–112.

[14] Hopcroft, J., and Karp, R. M., "An $n^{5/2}$ Algorithm for Maximum Matching in Bipartite Graphs", SIAM J. on Comput., 1975, pp. 225–231.

[15] Hall, P., "On Representation of Subsets", J. London Math. Soc. Vol. 10, 1935, pp. 26–30.

[16] Dilworth, R. P., "A Decomposition Theorem for Partially Ordered Sets," Ann. Math., Vol. 51, 1950, pp. 161–166.

Chapter 7

PLANAR GRAPHS

7.1 BRIDGES AND KURATOWSKI'S THEOREM

Consider a graph drawn in the plane in such a way that each vertex is represented by a point; each edge is represented by a continuous line connecting the two points which represent its end vertices and no two lines, which represent edges, share any points, except in their ends. Such a drawing is called a *plane graph*. If a graph G has a representation in the plane which is a plane graph then it is said to be *planar*.

In this chapter we shall discuss some of the classical work concerning planar graphs. The question of efficiently testing whether a given finite graph is planar will be discussed in the next chapter.

Let S be a set of vertices of a nonseparable graph $G(V, E)$. Consider the partition of the set V-S into classes, such that two vertices are in the same class if and only if there is a path connecting them which does not use any vertex of S. Each such class K defines a *component* as follows: The component is a subgraph $H(V', E')$ where $V' \supset K$. In addition, V' includes all the vertices of S which are connected by an edge to a vertex of K, in G. E' contains all edges of G which have at least one end vertex in K. An edge $u \overset{e}{-} v$, where both u and v are in S, defines a *singular component* $(\{u, v\}, \{e\})$. Clearly, two components share no edges, and the only vertices they can share are elements of S. The vertices of a component which are elements of S are called its *attachments*.

In our study we usually use a set S which is the set of vertices of a simple circuit C. In this case we call the components *bridges*; The edges of C are not considered bridges.

For example, consider the plane graph shown in Fig. 7.1. Let C be the outside boundary: $a \overset{1}{-} b \overset{2}{-} c \overset{3}{-} \cdots \overset{6}{-} g \overset{7}{-} a$. In this case the bridges are: $(\{e, g\}, \{8\})$, $(\{h, i, j, a, e, g\}, \{9, 10, 11, 12, 13, 14\})$, $(\{a, e\}, \{15\})$ and $(\{k, b, c, d\}, \{16, 17, 18\})$. The first and third bridges are singular.

Lemma 7.1: Let B be a bridge and a_1, a_2, a_3, three of its attachments. There exists a vertex v, not an attachment, for which there are three vertex

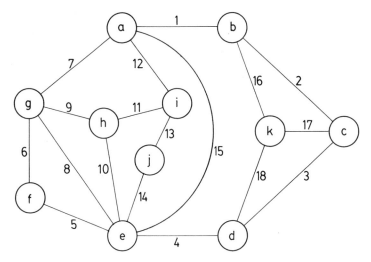

Figure 7.1

disjoint paths in B: $P_1(v, a_1)$, $P_2(v, a_2)$ and $P_3(v, a_3)$. ($P(a, b)$ denotes a path connecting a to b.)

Proof: Let $a_1 \overset{e_1}{-} v_1$, $a_2 \overset{e_2}{-} v_2$, $a_3 \overset{e_1}{-} v_3$ be edges of B. If any of the v_i's ($i = 1$, 2, 3) is an attachment then the corresponding edge is a singular bridge and is not part of B. Thus, $v_i \in K$, where K is the class which defines B. Hence, there is a simple path $P'(v_1, v_2)$ which uses vertices of K only; if $v_1 = v_2$, this path is empty. Also, there is a simple path $P''(v_3, v_1)$ which uses vertices of K only. Let v be the first vertex of $P''(v_3, v_1)$ which is also on P'. Now, let $P_1(v, a_1)$ be the part of P' which leads from v to v_1, concatenated with $v_1 - a_1$; $P_2(v, a_2)$ be the part of P' which leads from v to v_2, concatenated with $v_2 - a_2$; $P_3(v, a_3)$ be the part of P'' which leads from v to v_3, concatenated with $v_3 - a_3$. It is easy to see that these paths are disjoint.

Q.E.D.

Let C be a simple circuit of a nonseparable graph G, and $B_1, B_2, \ldots, B_k$ be the bridges with respect to C. We say that B_i and B_j *interlace* if at least one of the following conditions holds:

(i) There are two attachments of B_i, a and b, and two attachments of B_j, c and d, such that all four are distinct and they appear on C in the order a, c, b, d.
(ii) There are three attachments common to B_i and B_j.

For each bridge B_i, consider the subgraph $C + B_i$. If any of these graphs is not planar then clearly G is not planar. Now, assume all these subgraphs are planar. In every plane realization of G, C outlines a contour which divides the plane into two disjoint parts: its inside and outside. Each bridge must lie entirely in one of these parts. Clearly, if two bridges interlace they cannot be on the same side of C. Thus, in every plane realization of G the set of bridges is partitioned into two sets: those which are drawn inside C and those which are drawn outside. No two bridges in the same set interlace.

Lemma 7.2: If $B_1, B_2, \ldots, B_m$ is set of bridges of a nonseparable graph G with respect to a simple circuit C and the following two conditions are satisfied:

(1) For every $1 \le i \le n$, $C + B_i$ is planar,
(2) No two bridges interlace,
then $C + B_1 + B_2 + \cdots + B_m$ has a plane realization in which all these bridges are inside C.

Proof: We shall only outline the proof. As we go clockwise around C there must be a bridge B_i such that we encounter all its attachments in some order: $a_1, a_2, \ldots, a_t$, and no attachment of any other bridge appears between a_1 and a_t on C. Such a bridge can be found by starting from any attachment a of B_1 and going around, say clockwise. If before encountering all the attachments of B_1 we encounter an attachment of another bridge B_i, then all B_i's attachments are between two consecutive attachments of B_1 which may also belong to B_i. We repeat the same process on B_i, etc. Since the number of bridges is finite, and those discarded will not "interfere" with the new ones, the process will yield the desired bridge.

This observation allows a proof by induction. First B_i is drawn, and since no other bridge uses any of the vertices of C between a_1 and a_t, we can take C' to be the circuit which describes the part of C from a_t to a_1, clockwise, and the boundary of B_i, from a_1 to a_t, to form a simple circuit C', whose inside is so far empty. The remaining bridges are also bridges of C' and, clearly, satisfy (1) and (2) with respect to C'.

Q.E.D.

Corollary 7.1: Let G be a nonseparable graph and C a simple circuit in G. G is planar if and only if the bridges $B_1, B_2, \ldots, B_k$ of G, with respect to C, satisfy the following conditions:

(1) For every $1 \le i \le k$, $C + B_i$ is planar.

(2) The set of bridges can be partitioned into two subsets, such that no two bridges in the same subset interlace.

Let us introduce coloring of graphs. Here we consider only 2-coloring of vertices. In later chapters we shall discuss more general colorings. A graph $G(V, E)$ is said to be *2-colorable* if V can be partitioned into V_1 and V_2 in such a way that there is no edge in G with two end vertices in $V_1(V_2)$. (Obviously, a 2-colorable graph is bipartite, and vice versa. It is customary to use the term "bipartite" if the partition is given, and "2-colorable", if one exists.) The following theorem is due to König [1].

Theorem 7.1: A graph G is 2-colorable if and only if it has no odd length circuits.

Proof: It is easy to see that if a graph has an odd length circuit then it is not 2-colorable. In order to prove the converse, we may assume that G is connected; for if each component is 2-colorable then the whole graph is 2-colorable.

Let v be any vertex. Let us perform BFS (Section 1.5) starting from v. There cannot be an edge $u - w$ in G, if u and w belong to the same layer; i.e. are the same distance from v. For if such an edge exists then we can display an odd length circuit as follows: Let $P_1(v, u)$ be a shortest path from v to u. $P_2(v, w)$ is defined similarly and is of the same length. Let x be the last vertex which is common to P_1 and P_2. The part of P_1 from x to u, and the part of P_2 from x to w are of equal length, and together with $u - w$ they form an odd length simple circuit.

Now, we can color all the vertices of even distance from v with one color and all the vertices of odd distance from v with a second color.

$$Q.E.D.$$

We can use the concept of 2-colorability to decide whether the bridges B_1, $B_2, \ldots, B_k$, with respect to a simple circuit C can be partitioned into two subsets, each pairwise non interlacing, as follows: Construct a graph G' whose vertices are the bridges $B_1, B_2, \ldots, B_k$, and two vertices are connected by an edge if and only if the corresponding bridges in G interlace. Now test whether the graph G' is 2-colorable, by giving one vertex color 1 and using some search technique, such as DFS or BFS to color alternate vertices with different colors out of $\{1, 2\}$. If no contradiction arises, a coloring is obtained and thus, a partition. If a contradiction occurs, there is no 2-coloring, and therefore, no partition of the bridges; in this case the graph is nonplanar, by Corollary 7.1.

Let us now introduce the graphs of Kuratowski [2]: $K_{3,3}$ and K_5. They are shown in Fig. 7.2(a) and (b) respectively.

K_5 is a completely connected graph of 5 vertices, or a clique of 5 vertices. $K_{3,3}$ is a completely connected bipartite graph with 3 vertices on each side.

Lemma 7.3: Neither $K_{3,3}$ nor K_5 is planar

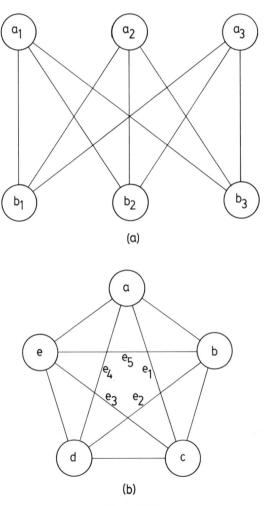

Figure 7.2

Proof: First, let us consider K_5, and its circuit C: $a - b - c - d - e - a$. Clearly, there are 5 bridges, all singular, corresponding to the edges e_1, e_2, e_3, e_4 and e_5. Let us construct G', as in the preceding discussion. It is shown in Fig. 7.3. For example, the bridge e_1 interlaces with e_5 and e_2, etc. Since G' contains an odd circuit, by Theorem 7.1 it is not 2-colorable, and the set of bridges of K_5 with respect to C is not partitionable. Thus, by Corollary 7.1, K_5 is not planar.

In the case of $K_{3,3}$, take C: $a_1 - b_1 - a_2 - b_2 - a_3 - b_3 - a_1$. The bridges $a_1 - b_2$, $a_2 - b_3$ and $a_3 - b_1$ form a triangle in the corresponding G'.

Q.E.D.

Before we take on Kuratowski's Theorem we need a few more definitions and a lemma.

Let $G(V, E)$ be a finite nonseparable plane graph with $|V| > 2$. A *face* of G is a maximal part of the plane such that for every two points x and y in it there is a continuous line from x and y which does not share with the realization of G any point. The contour of each face is a simple circuit of G; if any of these circuits is not simple then G is separable. Each of these circuits is called a *window*. One of the faces is of infinite area. It is called the *external* face, and its window is the external window. It is not hard to show that for every window W, there exists another plane realization, G', of the same graph, which has the same windows, but in G' W is external. First draw the graph on a sphere, maintaining the windows; this can be achieved by projecting

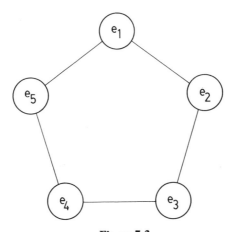

Figure 7.3

each point of the plane vertically up to the surface of a sphere whose center is in the plane and its intersecting circle with the plane encircles G. Next, place the sphere on a plane which is tangent to it, in such a way that a point in the face whose contour is W is the "north pole", i.e. furthest from the plane. Project each point P (other then the "north pole") of the sphere to the plane by a straight line which starts from the "north pole" and goes through P. The graph is now drawn in the plane and W is the external window.

Lemma 7.4: Let $G(V, E)$ be a 2-vertex connected graph with a separating pair a, b. Let $H_1, H_2, \ldots, H_m$ be the components with respect to $\{a, b\}$. G is planar if and only if for every $1 \le i \le m$ $H_i + (a \overset{e}{-} b)$ is planar, where $e \notin E$.

By $H_i + (a \overset{e}{-} b)$ we mean that a new edge connecting a and b is added to H_i.

Proof: In each H_i there is a path from a to b, or G is not 2-connected. Also $m > 1$. Thus, for each H_i we can find a path P from a to b in one of the other components. If G is planar, so is $H_i + P$, and therefore $H_i + (a \overset{e}{-} b)$ is planar. Now assume that each $H_i + (a \overset{e}{-} b)$ is planar. For each of these realization we can assume that e is on the external window. Thus, a planar realization of G exists, as demonstrated in Fig. 7.4 in the case of $m = 3$.

Q.E.D.

Two graphs are said to be *homeomorphic* if both can be obtained from the same graph by the insertion of new vertices of degree 2, in edges; i.e. an edge is replaced by a path whose intermediate vertices are all new.* Clearly, if two graphs are homeomorphic then either both are planar or both are not. We are now ready to state Kuratowski's Theorem [1].

Theorem 7.2: A graph G is nonplanar if and only if there is a subgraph of G which is homeomorphic to either $K_{3,3}$ or K_5.

Proof: If G has a subgraph H which is homeomorphic to either $K_{3,3}$ or K_5, then by Lemma 7.3, H is nonplanar, and therefore G is nonplanar. The converse is much harder to prove. We prove it by contradiction.

Let G be a graph which is nonplanar and which does not contain a

---

*Two graphs $G_1(V_1, E_1)$ and $G_2(V_2, E_2)$ are said to be **isomorphic** if there are 1-1 correspondences $f: V_1 \rightarrow V_2$ and $g: E_1 \rightarrow E_2$ such that for every edge $u \overset{e}{-} v$ in G_1 $f(u) \overset{g(e)}{-} f(v)$ in G_2. Clearly G_1 is planar if and only if G_2 is. Thus, we are not interested in the particular names of the vertices or edges, and distinguish between graphs only up to isomorphism.

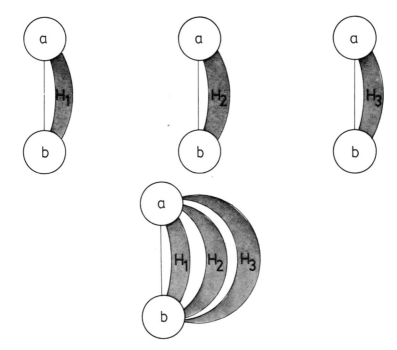

Figure 7.4

subgraph which is homeomorphic to one of Kuratowski's graphs; in addition let us assume that among such graphs G has the minimum number of edges.

First, let us show that the vertex connectivity of G is at least 3. Clearly, G is connected and nonseparable, or the number of its edges is not minimum. Assume it has a separating pair $\{a, b\}$, and let $H_1, H_2, \ldots, H_m$ be the components with respect to $\{a, b\}$, where $m > 1$. By Lemma 7.4, there exists an $1 \leq i \leq m$ for which $H_i + (a - b)$ is nonplanar. Clearly $H_i + (a - b)$ does not contain a subgraph which is homeomorphic to one of Kuratowski's graphs either. This contradicts the assumption that G has the minimum number of edges.

We now omit an edge $a_0 \overset{e_0}{-} b_0$ from G. The resulting graph, G_0, is planar. Since the connectivity of G is at least 3, G_0 is nonseparable. By Theorem 6.7, there is a simple circuit in G_0 which goes through a_0 and b_0. Let $\hat{G}_0$ be a plane realization of G_0 and C be a simple circuit which goes through a_0 and b_0, such that C encircles the maximum number of faces of all such circuits in all the plane realizations of G_0. Note that the selection of $\hat{G}_0$ and C is done simultaneously and not successively. Assuming u and v are vertices on C, let

$C[u, v]$ be the part of C going from u to v, clockwise. $C(u, v)$ is defined similarly, but the vertices u and v are excluded.

Consider now the external bridges of C in $\hat{G}_0$. If such a bridge, B, has two attachments either on $C[a_0, b_0]$ or $C[b_0, a_0]$ then C is not maximum. To see this, assume B has two attachments, a and b, in $C[a_0, b_0]$. There is a simple path $P(a, b)$ connecting a to b via edges and vertices of B, which is disjoint from C, and therefore is exterior to C. Form C' by adding to P the path $C[b, a]$. C' goes through a_0 and b_0, and the interior of C is either completely included in the interior of C' or in the exterior. In the first case C' has a larger interior than C, in $\hat{G}_0$. In the latter case we have to find another plane realization which is similar to G_0, but the exterior of C' is now the interior. In either case the maximality of the choice of $\hat{G}_0$ and C is contradictory.

By a similar argument we can also prove that neither a_0 nor b_0 can be an attachment of an external bridge. Thus, each bridge has one attachment in $C(a_0, b_0)$ and one in $C(b_0, a_0)$. Each of these bridges is singular (consists of a single edge); for otherwise its two attachments are a separating pair in G, contradicting its 3-connectivity. Finally, there is at least one such external singular bridge, or one could draw the edge e_0 outside C, to yield a planar realization of G. Similarly, there must be an internal bridge, B^*, which prevents the drawing of e_0 inside, and which cannot be transferred outside; i.e. B^* interlaces with an external singular bridge, say $a_1 \xrightarrow{e_1} b_1$. The situation is schematically shown in Fig. 7.5. We divide the argument to two cases according to whether B^* has any attachment other than a_0, b_0, a_1, b_1.

Case 1: B^* has an attachment a_2 other than a_0, b_0, a_1, b_1. Without loss of generality we may assume that a_2 is on $C(a_1, a_0)$. Since B^* prevents the drawing of e_0, it must have an attachment on $C(a_0, b_0)$. Since B^* interlaces e_1, it must have an attachment on $C(b_1, a_1)$.

Case 1.1: G^* has an attachment b_2 on $C(b_1, b_0)$. In B^*, there is a path P connecting a_2 with b_2. The situation is shown in Fig. 7.6. The subgraph of G shown in Fig. 7.6 is homeomorphic to $K_{3,3}$, where a_1, a_0 and b_2 play the role of the upper vertices of Fig. 7.2(a), and a_2, b_1 and b_0 play the role of the lower vertices.

Case 1.2: B^* has no attachment on $C(b_1, b_0)$. Thus, B^* has one attachment b_2' on $C(a_0, b_1]$; i.e., it may be b_1 but not a_0. Also, B^* has an attachment b_2'' on $C[b_0, a_1)$. By Lemma 7.1, there exists a vertex v and three vertex disjoint paths in B^*: $P_1(v, a_2)$, $P_2(v, b_2')$ and $P_3(v, b_2'')$. The situation is shown in Fig. 7.7. If we erase from the subgraph of G, shown in Fig. 7.7 the path $C[b_1, b_0]$ and all its intermediate vertices, the resulting subgraph is

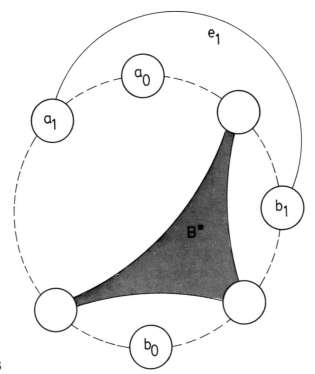

Figure 7.5

homeomorphic to $K_{3,3}$: Vertices a_2, b_2' and b_2'' play the role of the upper vertices, and a_0, a_1 and v, the lower vertices.

Case 2: B^* has no attachments other than a_0, b_0, a_1, b_1. In this case all four must be attachments; for if a_0 or b_0 are not, then B^* and e_1 do not interlace; if a_1 or b_1 are not, then B^* does not prevent the drawing of e_0.

Case 2.1 There is a vertex v, in B^*, from which there are four disjoint paths in B^*: $P_1(v, a_0)$, $P_2(v, b_0)$, $P_3(v, a_1)$ and $P_4(v, b_1)$. This case is shown in Fig. 7.8, and the shown subgraph is clearly homeomorphic to K_5.

Case 2.2: No vertex as in Case 2.1 exists. Let $P_0(a_0, b_0)$ and $P_1(a_1, b_1)$ be two simple paths in B^*. Let c_1 be the first vertex on P_1 which is common with P_0, and let c_2 be the last on P_1 which is common with P_0. We use only the first part, A, of P_1, connecting a_1 and c_1, and the last part, B, connecting c_2 with b_1. The pertaining subgraph of G is now shown in Fig. 7.9, and is

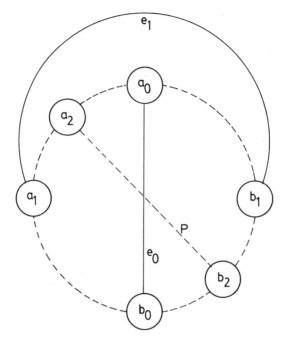

Figure 7.6

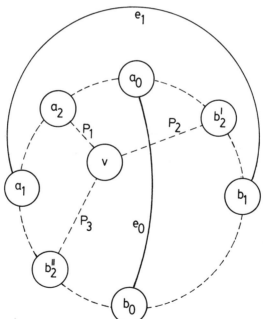

Figure 7.7

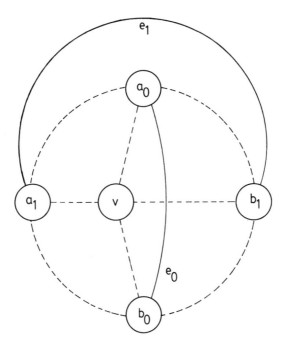

Figure 7.8

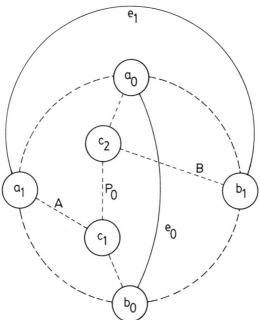

Figure 7.9

homeomorphic to $K_{3,3}$, after $C[a_0, b_1]$ and $C[b_0, a_1]$ and all their intermediate vertices are erased: Vertices a_0, b_1 and c_1 play the role of the upper vertices and b_0, a_1 and c_2, the lower. (If c_1 is closer to a_0 than c_2 then we erase $C[a_1, a_0]$ and $C[b_1, b_0]$, instead, and the upper vertices are a_0, a_1 and c_2.)

Q.E.D.

Kuratowski's theorem provides a necessary and sufficient condition for a graph to be planar. However, it does not yield an efficient algorithm for planarity testing. The obvious procedure, that of trying for all subsets of 5 vertices to see whether there are 10 vertex disjoint paths connecting all pairs, or for every pairs of 3 and 3 vertices whether there are 9 paths, suffers from two shortcomings. First there are $\binom{|V|}{5}$ choices of 5-sets and $\frac{1}{2} \binom{|V|}{3} \cdot \binom{|V|-3}{3}$ choices of 3 and 3 vertices; this alone is $O(|V|^6)$. But what is worse, we have no efficient way to look for the disjoint paths; this problem may, in fact, be exponential.

Fortunately, there are $O(|E|)$ tests, as we shall see in the next chapter, for testing whether a given graph is planar.

7.2 EQUIVALENCE

Let G_1 and $\hat{G}_2$ be two plane realizations of the graph G. We say that $\hat{G}_1$ and $\hat{G}_2$ are *equivalent* if every window of one of them is also a window in the other. G may be 2-connected and have nonequivalent plane realization; for example see Fig. 7.10.

Let us restrict our discussion to planar finite graphs, with no parallel edges and no self-loops, Our aim is to show that if the vertex connectivity of G, $c(G)$, is at least 3 then the plane realization of G is unique up to equivalence.

Lemma 7.5: A planar nonseparable graph G is 2-connected if there is a plane realization of it, $\hat{G}$, and one of its windows has more than one bridge.

Proof: If C is a window of $\hat{G}$ with more than one bridge, then all C's bridges are external. Therefore, no two bridges interlace. As in the first paragraph of the proof of Lemma 7.2, there exists a bridge B whose attachments can be ordered $a_1, a_2, \ldots, a_t$ and no attachments of any other bridge appear on $C(a_1, a_t)$. It is easy to see that $\{a_1, a_t\}$ is a separating pair; it separates the vertices of B and $C(a_1, a_t)$ from the set of vertices of all other bridges and $C(a_t, a_1)$, where neither set can be empty since G has no parallel edges.

Q.E.D.

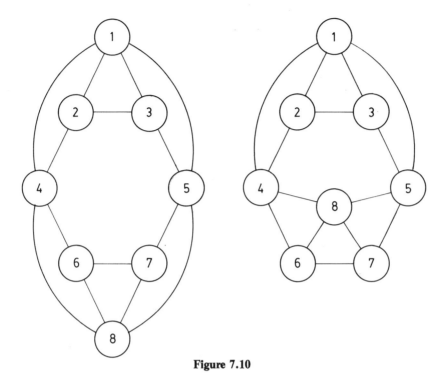

Figure 7.10

Theorem 7.3: If G is a plane graph with no parallel edges and no self-loops and if its vertex connectivity, $c(G)$, is at least 3, then every two plane realizations of G are equivalent.

Proof: Assume G has two plane realizations $\hat{G}_1$ and $\hat{G}_2$ which are not equivalent. Without loss of generality we may assume that there is a window C in $\hat{G}_1$ which is not a window in $\hat{G}_2$. Therefore, C has at least two bridges. By Lemma 7.5, and since C is a window in $\hat{G}_1$, G is 2-connected. A contradiction, since $c(G) \geq 3$.

Q.E.D.

7.3   EULER'S THEOREM

The following theorem is due to Euler.

Theorem 7.4: Let $G(V, E)$ be a non-empty connected plane graph. The number of faces, f, satisfies

$$|V| + f - |E| = 2. \tag{7.1}$$

Proof: By induction on $|E|$. If $|E| = 0$ then G consists of one vertex and there is one face, and 7.1 holds. Assume the theorem holds for all graphs with $m = |E|$. Let $G(V, E)$ be a connected plane graph with $m + 1$ edges. If G contains a circuit then we can remove one of its edges. The resulting plane graph is connected and has m edges, and therefore, by the inductive hypothesis, satisfies (7.1). Adding back the edge increases the number of faces by one and the number of edges by one, and thus (7.1) is maintained. If G contains no circuits, then it is a tree. By Lemma 2.1 it has at least two leaves. Removing a leaf and its incident edge yields a connected graph with one less edge and one less vertex which satisfies (7.1). Therefore, G satisfies (7.1) too.

Q.E.D.

The theorem implies that all connected plane graphs with $|V|$ vertices and $|E|$ edges have the same number of faces. There are many conclusions one can draw from the theorem. Some of them are the following:

Corollary 7.1: If $G(V, E)$ is a connected plane graph with no parallel edges, no self-loops and $|V| > 2$, then

$$|E| \leq 3|V| - 6. \tag{7.2}$$

Proof: Since there are no parallel edges, every window consists of at least three edges. Each edge appears on the windows of two faces, or twice on the window of one face. Thus, $3 \cdot f \leq 2 \cdot |E|$. By (7.1), $|E| = |V| + f - 2$. Thus, $|E| \leq |V| + \frac{2}{3}|E| - 2$, and (7.2) follows.

Q.E.D.

Corollary 7.2: Every connected plane graph with no parallel edges and no self-loops has at least one vertex whose degree is 5 or less.

Proof: Assume the contrary; i.e. the degree of every vertex is at least 6. Thus, $2 \cdot |E| \geq 6 \cdot |V|$; note that each edge is counted in each of its two end vertices. This contradicts (7.2).

Q.E.D.

7.4 DUALITY

Let $G(V, E)$ be a finite undirected and connected graph. A set $K \in E$ is called a *cutset* if it is a minimal separating set of edges; i.e. the removal of K

from G interrupts its connectivity, but no proper subset of K does it. It is easy to see that a cutset separates G into two connected components: Consider first the removal of $K - \{e\}$, where $e \in K$. G remains connected. Now remove e. Clearly G breaks into two components.

The graph $G_2(V_2, E_2)$ is said to be the *dual* of a connected graph $G_1(V_1, E_1)$ if there is a $1 - 1$ correspondence $f: E_1 \rightarrow E_2$, such that a set of edges S forms a simple circuit in G_1 if and only if $f(S)$ (the corresponding set of edges in G_2) forms a cutset in G_2. Consider the graph G_1 shown in Fig. 7.11(a). G_2 shown in Fig. 7.11(b) is a dual of G_1, but so is G_3, shown in Fig. 7.11(c), as the reader can verify by considering all (six) simple circuits of G_1 and all cusets of G_2, or G_3.

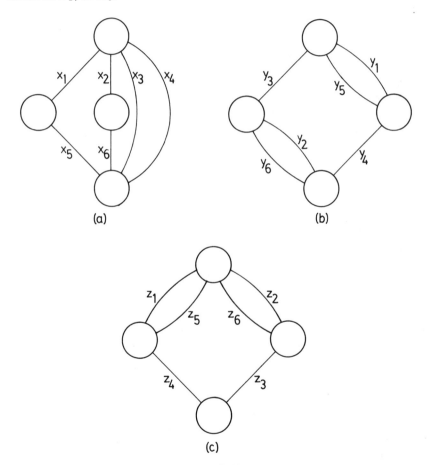

Figure 7.11

A *contraction* of an edge $x \overset{e}{-} y$ of a graph $G(V, E)$ is the following operation: Delete the edge e and merge x with y. The new contracted graph, G',
has one less edge and one less vertex, if $x \neq y$. Clearly, if G is connected so is
G'. A graph G' is a contraction of G if by repeated contractions we can construct G' from G.

Lemma 7.6: If a connected graph G_1 has a dual and G_1' is a connected
subgraph of G_1 then G_1' has a dual.

Proof: We can get G_1' from G_1 by a sequence of two kinds of deletions:

(i) A deletion of an edge e of the present graph, which is not in G_1', and
   whose deletion does not interrupt the connectivity of the present graph.
(ii) A deletion of a leaf of the present graph, which is not a vertex of G_1', together with its incident edge.

We want to show that each of the resulting graphs, starting with G_1 and
ending with G_1', has a dual.

Let G be one of these graphs, except the last, and its dual be G_d. First consider a deletion of type (i), of an edge e. Contract $f(e)$ in G_d, to get G_{dc}, If C
is a simple circuit in $G - e$, then clearly it cannot use e, and therefore it is a
circuit in G too. The set of edges of C is denoted by S. Thus, $f(S)$ is a cutset of
G_d, and it does not include $f(e)$. Thus, the end vertices of $f(e)$ are in the same
component of G_d with respect to $f(S)$. It follows that $f(S)$ is a cutset of G_{dc}
too. If K is a cutset of G_{dc} then it is a cutset of G_d too. Thus, $f^{-1}(K)$ form a
simple circuit C' in G. However, $f(e)$ is not in K, and therefore e is not in C'.
Hence, C' is a simple circuit of $G - e$.

Next, consider a deletion of type (ii) of a leaf v and its incident edge e.
Clearly, e, plays no role in a circuit. Thus, $f(e)$ cannot be a part of a cutset in
G_d. Hence, $f(e)$ is a self-loop. The deletion of v and e from G, and the contraction of $f(e)$ in G_d (which effectively, only deletes $f(e)$ from G_d), does not
change the sets of simple circuits in G and cutsets in G_d, and the correspondence is maintained.

Q.E.D.

Lemma 7.7: Let G be a connected graph and e_1, e_2 be two of its edges,
neither of which is a self loop. If for every cutset either both edges are in it or
both are not, then e_1 and e_2 are parallel edges.

Proof: If e_1 and e_2 are not parallel, then there exists a spanning tree which
includes both. (Such a tree can be found by contracting both edges and

finding a spanning tree of the contracted graph.) The edge e_1 is separating the tree into two connected components whose sets of vertices are S and $\bar{S}$. The set of edges between S and $\bar{S}$ in G in a cutset which includes e_1 and does not include e_2. A contradiction.

Q.E.D.

Lemma 7.8: Let G be a connected graph with a dual G_d and let f be the 1 − 1 correspondence of their edges. If $u \xrightarrow{e_1} x_1 \xrightarrow{e_2} x_2 \xrightarrow{e_3} \cdots x_{l-1} \xrightarrow{e_l} v$ is a simple path or circuit in G such that $x_1, x_2, \ldots, x_{l-1}$ are all of degree two, then $f(e_1), f(e_2), \ldots, f(e_l)$ are parallel edges in G_d.

Proof: Every circuit of G which contains one e_i, $1 \le i \le l$, contains all the rest. Thus, in G_d, if one edge, $f(e_i)$, is in a cutset then all the rest are. By Lemma 7.7, they are parallel edges.

Q.E.D.

Lemma 7.9: If a connected graph G_1 has a dual and G_2 is homeomorphic to G_1 then G_2 has a dual.

Proof: If an edge $x \xrightarrow{e} y$ of G_1 is replaced in G_2 by a path $x \xrightarrow{e_1} v_1 \xrightarrow{e_2} v_2 - \cdots \xrightarrow{e_l} y$, then in the dual $f(e)$ is replaced by parallel edges: $f(e_1), f(e_2), \ldots, f(e_l)$. If a path of G_1 is replaced in G_2 by an edge e, then the edges of the dual which correspond to the edges of the path are all parallel (by Lemma 7.8) and can be replaced by a single edge $f(e)$. It is easy to see that every circuit which uses an edge e, when it is replaced by a path, will use all the edges of the path instead; while in the dual, every cutset which uses $f(e)$ will use all the parallel edges which replace it. Thus, the correspondence of circuits and cutsets is maintained.

Q.E.D.

Theorem 7.5: A (connected) graph has a dual if and only if it is planar.

Proof: Assume $G_1(V_1, E_1)$ is a planar graph and $\hat{G}_1$ is a plane realization of it. Choose a point p_i in each face F_i of $\hat{G}_1$. Let V_2 be the set of these points. Since G_1 is connected, the boundary of every face is a circuit (not necessarily simple) which we call its window. Let W_i be the window of F_i, and assume it consists of l edges. We can find l lines, all starting in p_i, but no two share any other point, such that each of the lines ends on one of the l edges of W_i, one line per edge. If a separating edge* e appears on W_i, then clearly it appears

---

*An edge whose removal from G_1 interrupts its connectivity.

twice. In this case there will be two lines from p_i hitting e from both directions. These two lines can be made to end in the same point on e, thus creating a closed curve which goes through p_i and crosses e. If e is not a separating edge then it appears on two windows, say W_i and W_j. In this case we can make the line from p_i to e meet e at the same point as does the line from p_j to e, to form a line connecting p_i with p_j which crosses e. None of the set of lines, thus formed crosses another, and we have one per edge of $\hat{G}_1$. Now define $\hat{G}_2(V_2. E_2)$ as follows: The set of lines connecting the p_i's is a representation of E_2. The 1-1 correspondence $f\colon E_1 \rightarrow E_2$ is defined as follows: $f(e)$ is the edge of $\hat{G}_2$ which crosses e. Clearly, $\hat{G}_2$ is a plane graph which is a realization of a graph $G_2(V_2, E_2)$. It remains to show that there is a 1-1 correspondence of the simple circuits of G_1 to the cutsets of G_2. The construction described above is demonstrated in Fig. 7.12, where $\hat{G}_1$ is shown in solid lines, and $\hat{G}_2$ is shown in dashed lines.

Let C be a simple circuit of G_1. Clearly, in $\hat{G}_1$, C describes a simple closed curve in the plane. There must be at least one vertex of $\hat{G}_2$ inside this circuit, since at least one edge of $\hat{G}_2$ crosses the circuit, and it crosses the circuit exactly once. The same argument applies to the outside too. This implies that $f(S)$, where S is the set of edges of C, forms a separating set of G_2. Let us postpone the proof of the minimality of $f(S)$ for a little while.

Now, let K be a cutset of G_2. Let T and $\bar{T}$ be the sets of vertices of the two

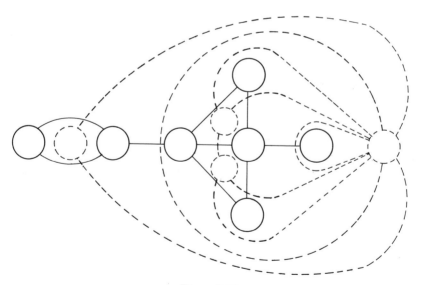

Figure 7.12

components of G_2 formed by the deletion of K. The set of faces of $\hat{G}_1$ which correspond to the vertices of T, forms a continuous region, but not the whole plane. The minimality of K implies that the boundary of this region is a circuit in G_1, whose edges correspond to K. Thus we have shown that every simple circuit of G_1 corresponds to a separating set of G_2, and every cutset of G_2 corresponds to a circuit of G_1.

Now let us handle the minimality. If S is the set of edges of a simple circuit C of G_1 and $f(S)$ is not a cutset of G_2, then there is a proper subset of $f(S)$ which is a cutset, say K. Therefore, $f^{-1}(K)$ is a circuit of G_1. However $f^{-1}(K) \subsetneq S$. A contradiction to the assumption that C is simple. The proof that if K is a cutset then $f^{-1}(K)$ is a simple circuit is similar.

This completes the proof that a connected planar graph has a dual. We turn to the proof that a nonplanar graph has no dual. Here we follow the proof of Parson [3].

First let us show that neither $K_{3,3}$ nor K_5 have a dual. In $K_{3,3}$ the shortest circuit is of length 4, and for every two edges there exists a simple circuit which uses one but does not use the other. Thus, in its dual no cutset consists of less than 4 edges and there are no parallel edges. Therefore, the degree of each vertex is at least 4 and there must be at least 5 vertices. The number of edges is, therefore, at least 10 $(5 \cdot 4/2)$, while $K_{3,3}$ has 9 edges. Thus, $K_{3,3}$ has no dual. In the case of K_5, it has 10 edges, 10 simple circuits of length 3 and no shorter circuits, and for every two edges there is a simple circuit which uses one but not the other. Thus, the dual must have 10 edges, 10 cutsets of three edges, no cutset of lesser size and therefore the degree of every vertex is at least three. Also, there are no parallel edges in the dual. If the dual has 5 vertices then it is K_5 itself (10 edges and no parallel edges), but K_5 has no cutsets of three edges. If the dual has 7 vertices or more, then it has at least 11 edges ($\lceil (7 \cdot 3/2 \rceil$). Thus, the dual must have 6 vertices. Since it has 10 custsets of 3 edges, there is one which separates S from $\bar{S}$ where neither consists of a single vertex. If $|S| = 2$ then it contains a vertex whose degree is 2. If $|S| = |\bar{S}| = 3$, then the maximum number of edges in the dual is 9. Thus, K_5 has no dual.

Now, if G is a nonplanar graph with a dual, then by Kuratowski's theorem (Theorem 7.2) it contains a subgraph G' which is homeomorphic to either $K_{3,3}$ or K_5. By Lemma 7.6, G' has a dual too. By Lemma 7.9, either $K_{3,3}$ or K_5 have a dual. A contradiction. Thus, no nonplanar graph has a dual.

$$\text{Q.E.D.}$$

Theorem 7.5 provides another necessary and sufficient condition for planarity, in addition to Kuratowski's theorem. However, neither has been shown to be useful for testing planarity.

There are many facts about duality which we have not discussed. Among them are the following:

(1) If G_d is a dual of G then G is a dual of G_d.
(2) A 3-connected planar graph has a unique dual.

The interested reader can find additional information and references in the books of Harary [4], Ore [5], and Wilson [6].

PROBLEMS

7.1 The purpose of this problem is to prove a variation of Kuratowski's theorem.

   (a) Prove that if G is a connected graph and v_1, v_2, v_3 are three vertices then there exists a vertex v and three (vertex) disjoint paths $P_1(v, v_1)$, $P_2(v, v_2)$ and $P_3(v, v_3)$, one of which may be empty.
   (b) Prove that if G is a connected graph and S is a set of four vertices then either there is a vertex v with four disjoint paths to the members of S or there are two vertices u and v such that two members of S are connected to u by paths, two to v, and u is connected to v; all five paths are vertex disjoint.
   (c) Show that if a graph is contractible to $K_{3,3}$ then it has a subgraph homeomorphic to $K_{3,3}$.
   (d) Show that if a graph is contractible to K_5 then either it has a subgraph which is homeomorphic to K_5 or it has a subgraph which is contractible to $K_{3,3}$.
   (e) Prove the theorem: A graph is nonplanar if and only if it contains a subgraph which is contractible to $K_{3,3}$ or K_5.

7.2 Show a graph which is nonplanar but is not contractible to either $K_{3,3}$ or K_5. Does it contradict the result of Problem 7.1(e)?

7.3 Use Kuratowski's theorem to prove that the Petersen graph, shown below is nonplanar.

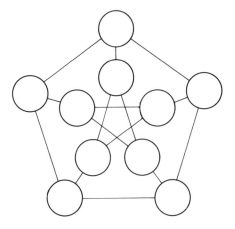

7.4 Is the graph shown below planar? Justify your answer.

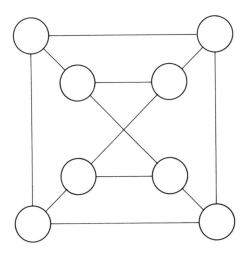

7.5 A plane graph is called *triangular* if each of its windows is a triangle. Let N_i be the number of vertices whose degree is i.

(a) Prove that every plane graph with 3 or more vertices which has no self-loops and no parallel edges can be made triangular, by adding new edges without creating any self-loops or parallel edges. (This process is called *triangulation.*)

(b) Let G be a triangular plane graph as above, with $|V| > 3$. Prove that

$$N_1 = N_2 = 0.$$

(c) Let G be a triangular plane graph as above. Prove that

$$12 = 3 \cdot N_3 + 2 \cdot N_4 + N_5 - N_7 - 2 \cdot N_8 - 3 \cdot N_9 - 4 \cdot N_{10} \cdots$$

(d) Prove that in a graph, as above, if there are no vertices of degree 3 or 4 then there are at least 12 vertices of degree 5.

(e) Prove that if G is a planar graph with no self-loops or parallel edges, and $|V| > 3$, then it has at least 4 vertices with degrees less than 6 and if $N_3 = N_4 = 0$ then $N_5 \geq 12$.

(f) Prove that if the vertex connectivity, $c(G)$, of a graph $G(V, E)$ is at least 5 then $|V| \geq 12$.

(g) Prove that if G is planar then $c(G) < 6$.

7.6 Prove that if $G_1(V_1, E_1)$ and $G_2(V_2, E_2)$ are homeomorphic then

$$|E_1| - |V_1| = |E_2| - |V_2|.$$

7.7 Show a triangular plane graph, without parallel edges, which is not hamiltonian.

Let G be a plane graph. The plane graph G^*, which is constructed by the procedure in the first part of the proof of Theorem 7.5, is called its *geometric dual.*

7.8 Prove that if G^* is the geometric dual of G then G is the geometric dual of G^*.

REFERENCES

[1] König, D., *Theorie der endlichen und unendlichen Graphen*. Leipzig, 1936. Reprinted Chelsea, 1950.

[2] Kuratowski, K., "Sur le Problème des Courbes Gauches en Topologie", Fund. Math., Vol. 15, 1930, pp. 217–283.

[3] Parson, T. D., "On Planar Graphs", Am. Math. Monthly, Vol. 78, No. 2, 1971, pp. 176–178.

[4] Harary, F., *Graph Theory*, Addison Wesley, 1969.

[5] Ore, O., *The Four Color Problem*, Academic Press, 1967.

[6] Wilson, R. J., *Intr. to Graph Theory*, Longman, 1972.

Chapter 8

TESTING GRAPH PLANARITY

8.1 Introduction

There are two known planarity testing algorithms which have been shown to be realizable in a way which achieves linear time ($O(|V|)$). The idea in both is to follow the decisions to be made during the planar construction of the graph, piece by piece, as to the relative location of the various pieces. The construction is not carried out explicitly because there are difficulties, such as crowding of elements into a relatively small portion of the area allocated, that, as yet, we do not know to avoid. Also, an explicit drawing of the graph is not necessary, as we shall see, to decide whether such a drawing is possible. We shall imagine that such a realization is being carried out, but will only decide where the various pieces are laid, relative to each other, and not of their exact shape. Such decisions may change later in order to make place for later additions of pieces. In both cases it was shown that the algorithm terminates within $O(|V|)$ steps, and if it fails to find a "realization" then none exists.

The first algorithm starts by finding a simple circuit and adding to it one simple path at a time. Each such new path connects two old vertices via new edges and vertices. (Whole pieces are sometimes flipped over, around some line). Thus, we call it the *path addition* algorithm. The basic ideas were suggested by various authors, such as Auslander and Parter [1] and Goldstein [2], but the algorithm in its present form, both from the graph theoretic point of view, and complexity point of view, is the contribution of Hopcroft and Tarjan [3]. They were first to show that planarity testing can be done in linear time.

The second algorithm adds in each step one vertex. Previously drawn edges incident to this vertex are connected to it, and new edges incident to it are drawn and their other endpoints are left unconnected. (Here too, sometimes whole pieces have to be flipped around or permuted). The algorithm is due to Lempel, Even and Cederbaum [4]. It consists of two parts. The first part was shown to be linarily realizable by Even and Tarjan [5];

the second part was shown to be linearly realizable by Leuker and Booth [6]. We call this algorithm the *vertex addition* algorithm.

Each of these algorithms can be divided into its graph theoretic part and its data structures and their manipulation. The algorithms are fairly complex and a complete description and proof would require a much more elaborate exposition. Thus, since this is a book on graphs and not on programming, I have chosen to describe in full the graph theoretic aspects of both algorithms, and only briefly describe the details of the data manipulation techniques. An attempt is made to convince the reader that the algorithms work, but in order to see the details which make it linear he will have to refer to the papers mentioned above.

Throughout this chapter, for reasons explained in Chapter 7, we shall assume that $G(V, E)$ is a finite undirected graph with no parallel edges and no self loops. Also, we shall assume that G is nonseparable. The first thing that we can do is check whether $|E| \leq 3 \cdot |V| - 6$. By Carollary 7.1, if this condition does not hold then G is nonplanar. Thus, we can restrict our algorithms to the cases where $|E| = O(|V|)$.

8.2   The Path Addition Algorithm of Hopcroft and Tarjan

The algorithm starts with a DFS of G. We assume that G is nonseparable. Thus, we drop from the DFS the steps for testing nonseparability. However we still shall need the lowpoint function, to be denoted now $L1(v)$. In addition we shall need the *second lowpoint* function, $L2(v)$, to be defined as follows. Let $S(v)$ be the set of values $k(u)$ of vertices u reachable from descendants of v by a single back edge. Clearly, $L1(v) = \text{Min} \{\{k(v)\} \cup S(v)\}$.

Define

$$L2(v) = \text{Min} \{\{k(v)\} \cup [S(v) - \{L1(v)\}]\}. \tag{8.1}$$

Let us now rewrite the DFS in order to compute these functions:

(1) Mark all the edges "unused". For every $v \in V$ let $k(v) \leftarrow 0$. Let $i \leftarrow 0$ and $v \leftarrow s$ (the vertex s is where we choose to start the DFS).

(2) $i \leftarrow i + 1$, $k(v) \leftarrow i$, $L1(v) \leftarrow i$, $L2(v) \leftarrow i$.

(3) If v has no unused incident edges go to Step (5).

(4) Choose an unused incident edge $v \xrightarrow{e} u$. Mark e "used". If $k(u) \neq 0$ then do the following:

If $k(u) < L1(v)$ then $L2(v) \leftarrow L1(v)$,
$$L1(v) \leftarrow k(u).$$
If $k(u) > L1(v)$ then $L2(v) \leftarrow \text{Min } \{L2(v), k(u)\}$.
Go to Step (3).
Otherwise $(k(u) = 0)$, let $f(u) \leftarrow v$, $v \leftarrow u$ and go to Step (2).
(5) If $k(v) = 1$, Halt.
(6) $(k(v) > 1$; we backtrack).
If $L1(v) < L1(f(v))$ then $L2(f(v)) \leftarrow \text{Min } \{L2(v), L1(f(v))\}$,
$$L1(f(v)) \leftarrow L1(v).$$
If $L1(v) = L1(f(v))$ then $L2(f(v)) \leftarrow \text{Min } \{L2(v), L2(f(v))\}$.
Otherwise $(L1(v) > L1(f(v)))$, $L2(f(v)) \leftarrow \text{Min } \{L1(v), L2(f(v))\}$.
$v \leftarrow f(v)$.
Go to Step (3).

From now on, we refer to the vertices by their $k(v)$ number; i.e. we change the name of v to $k(v)$.

Let $A(v)$ be the *adjacency list* of v; i.e. the list of edges incident from v. We remind the reader that after the DFS each of the edges is directed; the tree edges are directed from low to high and the back edges are directed from high to low. Thus, each edge appears once in the adjacency lists. Now, we want to reorder the adjacency lists, but first, we must define an order on the edges. Let the value $\phi(e)$ of an edge $u \xrightarrow{e} v$ be defined as follows:

$$\phi(e) = \begin{cases} 2 \cdot v & \text{if } u \xrightarrow{e} v \text{ is a back edge.} \\ 2 \cdot L1(v) & \text{if } u \xrightarrow{e} v \text{ is a tree edge and } L2(v) \geq u. \\ 2 \cdot L1(v) + 1 & \text{if } u \xrightarrow{e} v \text{ is a tree edge and } L2(v) < u. \end{cases}$$

Next we order the edges in each adjacency list to be in nondecreasing order with respect to ϕ. [This can be done in $O(|V|)$ time as follows. First compute for each edge its ϕ value. Prepare $2 \cdot |V| + 1$ buckets, numbered $1, 2, \ldots, 2 \cdot |V| + 1$. Put each e into the $\phi(e)$ bucket. Now, empty the buckets in order, first bucket number 1, then 2 etc., putting the edges taken out into the proper new adjacency lists, in the order that they are taken out of the buckets.]

The new adjacency lists are now used, in a second run of a DFS algorithm, to generate the paths, one by one. In this second DFS, vertices are not renumbered and there is no need to recompute $f(v)$, $L1(v)$ or $L2(v)$. The tree remains the same, although the vertices may not be visited in the same order. The paths finding algorithm is as follows:

(1) Mark all edges "unused" and let $v \leftarrow 1$.
(2) Start the circuit C; its first vertex is 1.
(3) Let the first edge on $A(v)$ be $v \xrightarrow{e} u$.
    If $u \neq 1$ then add u to C, $v \leftarrow u$ and repeat Step (3). Otherwise, $(u = 1)$, C is closed. Output C.
(4) If $v = 1$, halt.
(5) If $A(v)$ is used up then $v \leftarrow f(v)$ and go to Step (4).
(6) Start a path P with v as its first vertex.
(7) Let $v \xrightarrow{e} u$ be the first unused edge on $A(v)$.
    If e is a back edge $(u < v)$ terminate P with u, output P and go to (5).
(8) (e is a tree edge). Put u on P, $v \leftarrow u$ and go to Step (7).

Lemma 8.1: The paths finding algorithm finds first a circuit C which consists of a path from 1 (the root) to some vertex v, via tree edges, and a back edge from v to 1.

Proof: Let $1 \rightarrow u$ be the first edge of the tree to be traced (in the first application of Step (3)). We assume that G is nonseparable and $|V| > 2$. Thus, by Lemma 3.7, this edge is the only tree edge out of 1, and $u = 2$. Also, 2 has some descendants, other than itself. Clearly, $2 \rightarrow 3$ is a tree edge. By Lemma 3.5, $L1(3) < 2$, i.e. $L1(3) = 1$. Thus $L1(2) = 1$. The reordering of the adjacency lists assures that the first path to be chosen out of 1 will lead back to 1 as claimed.

Q.E.D.

Lemma 8.2: Each generated path P is simple and it contains exactly two vertices in common with previously generated paths; they are the first vertex, f, and the last l.

Proof: The edge scanning during the paths finding algorithm is in a DFS manner, in accord with the structure of the tree (but not necessarily in the same scanning order of vertices). Thus, a path starts from some (old) vertex f, goes along tree edges, via intermediate vertices which are all new, and ends with a back edge which leads to l. Since back edges always lead to ancestors, l is old. Also, by the reordering of the adjacency lists and the assumption that G is nonseparable l must be lower than f. Thus, the path is simple.

Q.E.D.

Let S_v denote the set of descendants of v, including v itself.

Lemma 8.3: Let f and l be the first and last vertices of a generated path P and $f \to v$ be its first edge.
(i) if $v \neq l$ then $L1(v) = l$.
(ii) l is the lowest vertex reachable from S_f via a back edge which has not been used in any path yet.

Proof: Let us partition $S_f - \{f\}$ into two parts, α and β, as follows. A descendant u belongs to α if and only if when the construction of P begins, we have already backtracked from u. Clearly, $f \in \beta$ and all the back edges out of α have been scanned already. Let u be the lowest vertex reachable via an unused back edge from β. Clearly, the first remaining (unused) edge of $A(f)$ is the beginning of a directed path to u, which is either an unused back edge from f to u or a path of tree edges, via vertices of β, followed by a single back edge to u. Thus, $u = l$, and the Lemma follows.

Q.E.D.

Lemma 8.4: Let P_1 and P_2 be two generated paths whose first and last vertices are f_1, l_1 and f_2, l_2, respectively. If P_1 is generated before P_2 and f_2 is a descendant of f_1 then $l_1 \leq l_2$.

Proof: The Lemma follows immediately from Lemma 8.3.

Q.E.D.

So far, we have not made any use of $L2(v)$. However, the following lemma relies on it.

Lemma 8.5: Let $P_1 : f \xrightarrow{1} v_1 \to \cdots \to l$ and $P_2 : f \xrightarrow{e_2} v_2 \to \cdots \to l$ be two generated paths, where P_1 is generated before P_2. If $v_1 \neq l$ and $L2(v_1) < f$ then $v_2 \neq l$ and $L2(v_2) < f$.

Proof: By the definition of $\phi(e)$, $\phi(e_1) = 2 \cdot l + 1$. If $v_2 = l$ or $L2(v_2) \geq f$ then $\phi(e_2) = 2 \cdot l$, and e_2 should appear in $A(f)$ before e_1. A contradiction.

Q.E.D.

Let C be $1 \to v_1 \to v_2 \to \cdots \to v_n \to 1$. Clearly $1 < v_1 < v_2 \cdots < v_n$. Consider now the bridges of G with respect to C.

Lemma 8.6: Let B be a non-singular bridge of G with respect to C, whose highest attachment is v_i. There exist an tree edge $v_i \xrightarrow{*} u$ which belongs to B and all other edges of B with endpoints on C are back edges.

Proof: The Lemma follows from the fact that the paths finding algorithm is a DFS. First C is found. We then backtrack from a vertex v_j only if all its descendants have been scanned. No internal part of B can be scanned before we backtrack into v_i. There must be a tree edge $v_i \rightarrow u$, where u belongs to B, for the following reasons. If all the edges of B, incident to v_i are back edges, they all must come from descendents or go to ancestors of v_i (see Lemma 3.4). An edge from v_i to one of its ancestor (which must be on C) is a singular bridge and is not part of B. An edge from a descendant w of v_i to v_i implies that w cannot be in B, for it has been scanned already, and we have observed that no internal part of B can be scanned before we backtrack into v_i. If any other edge $v_k - x$ of B is also a tree edge then, by the definition of a bridge, there is a path connecting u and x which is vertex disjoint from C. Along this path there is at least one edge which contradicts Lemma 3.4.

Q.E.D.

Corollary 8.1: Once a bridge B is entered, it is completely traced before it is left.

Proof: By Lemma 8.6, there is only one edge through which B is entered. Since eventually the whole graph is scanned, and no edge is scanned twice in the same direction, the corollary follows.

Q.E.D.

Assuming C and the bridges explored from vertices higher than v_i have already been explored and drawn in the plane. The following lemma provides a test for whether the next generated path could be drawn inside; the answer is negative even if the path itself can be placed inside, but it is already clear that the whole bridge to which it belongs cannot be placed there.

Lemma 8.7: Let $v_i \rightarrow u_1 \rightarrow u_2 \rightarrow \cdots \rightarrow u_l(=v_j)$ be the first path, P, of the bridge B to be generated by the paths finding algorithm. P can be drawn inside (outside) C if there is no back edge $w \rightarrow v_k$ drawn inside (outside) for which $j < k < i$. If there is such an edge, then B cannot be drawn inside (outside).

Proof: The sufficiency is immediate. If there is no back edge, drawn inside C, which enters the path $v_j \rightarrow v_{j+1} \rightarrow \cdots v_i$ in one of its internal vertices, then there cannot be any inside edge incident to these vertices, since bridges to be explored from vertices lower than v_i have not been scanned yet. Thus, P can be drawn inside if it is placed sufficiently close to C.

Now, assume there is a back edge $w \to v_k$, drawn inside, for which $j < k < i$. Let P' be the directed path from v_p to v_k whose last edge is the back edge $w \to v_k$. Clearly v_p is on C and $p \geq i$; P' is not necessarily generated in one piece by the path finding algorithm, if it is not the first path to be generated in the bridge B' to which it belongs.

Case 1: $p > i$. The bridges B and B' interlace by part (i) of the definition of interlacement. Thus, B cannot be drawn inside.

Case 2: $p = i$. Let P'' be the first path of B' to be generated, P'': $v_i \to x_1 \to x_2 \to \cdots \to v_q$. By Lemma 8.4, $q \leq j$, since B' is explored before B.

Case 2.1: $q < j$. Since v_i and v_j are attachments of B, v_k and v_q are attachments of B' and $q < j < k < i$, the two bridges interlace. Thus, B cannot be drawn inside.

Case 2.2: $q = j$. P'' cannot consist of a single edge, for in this case it is a singular bridge and v_k is not one of its attachments. Also, $L2(x_1) \leq v_k$. Thus, $L2(x_1) < v_i$. By Lemma 8.5, $u_1 \neq v_j$ and $L2(u_1) < v_i$. This implies that B and B' interlace by either part (i) or part (ii) of the definition of interlacement, and B cannot be drawn inside.

$$\text{Q.E.D.}$$

The algorithm assumes that the first path of the new bridge B is drawn inside C. Now, we use the results of Corollary 7.1, Theorem 7.1 and the discussion which follows it, to decide whether the part of the graph explored so far is planar, assuming that $C + B$ is planar. By Lemma 8.7, we find which previous bridges interlace with B. The part of the graph explored so far is planar if and only if the set of its bridges can be partitioned into two sets such that no two bridges in the same set interlace. If the answer is negative, the algorithm halts declaring the graph nonplanar. If the answer is positive, we still have to check whether $C + B$ is planar.

Let the first path of B be P: $v_i \to u_1 \to u_2 \to \cdots \to v_j$. We now have a circuit C' consisting of $C[v_j, v_i]$ and P. The rest of C is an outside path P', with respect to C', and it consists of $C[v_i, 1]$ and $C[1, v_j]$. The graph $B + C[v_j, v_i]$ may have bridges with respect to C', but none of them has all its attachments on $C[v_j, v_i]$, for such a bridge is also a bridge of G with respect to C, and is not a part of B.

Thus, no bridge of C', with attachments on $C(v_j, v_i)$ may be drawn outside C', since it interlaces with P'. We conclude that $C + B$ is planar if and only if $B + C[v_j, v_i]$ is planar and its bridges can be partitioned into an inside set and an outside set so that the outside set contains no bridge with attachments in $C(v_j, v_i)$. The planarity of $B + C[v_j, v_i]$ is tested by applying the algorithm recursively, using the established vertex numbering $L1$, $L2$, f functions and ordering of the adjacency lists. If $B + C[v_j, v_i]$ is found to be nonplanar, clearly G is nonplanar. If it is found to be planar, we check whether all its bridges with attachments in $C(v_j, v_i)$ can be placed inside. If so, $C + B$ is planar; if not, G is nonplanar.

Hopcroft and Tarjan devised a simple tool for deciding whether the set of bridges can be partitioned properly. It consists of three stacks Π_1, Π_2 and Π_3. Π_1 contains in nondecreasing order (the greatest entry on top) the vertices of $C(1, v_i)$ (where v_i is the last vertex of C into which we have backtraced), which are attachments of bridges placed inside C in the present partition of the bridges. A vertex may appear on Π_1 several times, if there are several back edges into it from inside bridges. Π_2 is similarly defined for the outside bridges. Both Π_1 and Π_2 are doubly linked lists, in order to enable switching over of complete sections from one of them to the other, investing, per section, time bounded by some constant. Π_3 consists of pairs of pointers to entries in Π_1 and Π_2. Its task will be described shortly.

Let S be a maximal set of bridges, explored up to now, such that a decision for any one of these bridges as to the side it is in, implies a decision for all the rest. (S corresponds to a connected component of the graph of bridges, as in the discussion following Theorem 7.1.) Let the set of entries in Π_1 and Π_2, which correspond to back edges from the bridges of S, be called a *block*.

Lemma 8.8: Let K be a block, whose highest element is v_h and lowest element is v_l. If v_p is an entry in Π_1 or Π_2 then v_p belongs to K if $v_l < v_p < v_h$.

Proof: We prove the lemma by induction on the order in which the bridges are explored. At first, both Π_1 and Π_2 are empty and the lemma is vacuously true. The lemma trivially holds after one bridge is explored.

Assume the lemma is true up to the exploration of the first path P of the new bridge B, where $P: v_i \rightarrow \ldots \rightarrow v_j$. If there is no vertex v_k on Π_1 or Π_2 such that $v_j < v_k < v_i$ then clearly the attachments of B (in $C(1, v_i)$) form a new block (assuming $C + B$ is planar) and the lemma holds. However, if there are vertices of Π_1 or Π_2 in between v_j and v_i then, by Lemma 8.7, the bridges, they are attachments of, all interlace with B. Thus, the old blocks which these attachments belong to, must now be merged into one new block with the attachments of B (in $C(1, v_i)$). Now, let v_l be the lowest vertex of the new block and v_h be the highest. Clearly, v_l was the lowest vertex of some old block whose highest vertex was v_h', and $v_h' > v_j$. Thus, by the inductive hypothesis, no attachment of another block could be in between v_l and v_h', and therefore cannot be in this region after the merger. Also, all the attachments in between v_j and v_h are in the new block since they are attachments of bridges which interlace with B. Thus, all the entries of Π_1 or Π_2 which are in between v_l and v_h belong to the new block.

Q.E.D.

Corollary 8.2: The entries of one block appear consecutively on $\Pi_1(\Pi_2)$.

Thus, when we consider the first path P: $v_i \rightarrow \cdots \rightarrow v_j$ of the new bridge B, in order to decide whether it can be drawn inside, we check the top entries t_1 and t_2 of Π_1 and Π_2 respectively.

If $v_j \geq t_1$ and $v_j \geq t_2$ then no merger is necessary; the attachments of B (in $C(1, v_i)$) are entered as a block (if $C + B$ is found to be planar) on top of Π_1.

If $v_j < t_1$ and $v_j \geq t_2$ then we first join all the blocks for which their highest entry is higher than v_j. To this end there is no need to check Π_2, since $v_j \geq t_2$, but still several blocks may merge. Next, switch the sections of the new block, i.e. the section on Π_1 exchanges places with the section on Π_2. Finally, place the attachments of B in nondecreasing order, on top of Π_1; these entries join the new block.

If $v_j \geq t_1$ and $v_j < t_2$ then again we join all the blocks whose highest element is greater than v_j; only the sections on Π_2 need to be checked. The attachments of B join the same block and are placed on top of Π_1.

If $v_j < t_1$ and $v_j < t_2$ then all blocks whose highest element is greater than v_j are joined into one new block. As we join the blocks we examine them one by one. If the highest entry in the section on Π_1 is higher than v_j then we switch the sections. If it is still higher, then we halt and declare the graph nonplanar. If all these switches succeed, then the merger is completed by adding the attachments of B on top of Π_1.

In order to handle the sections switching, the examination of their tops and mergers efficiently, we use a third stack, Π_3. It consists of pairs of pointers, one pair (x, y) per block; x points to the lowest entry of the section in Π_1 and y to the lowest entry of the section in Π_2. (If the section is empty then the pointer's value is 0). Two adjacent blocks are joined by simply discarding the pair of the top one. When several blocks have to be joined together upon the drawing of the first path of a new bridge, only the pair of the lowest of these blocks need remain, except when one of its entries is 0. In this case the lowest nonzero entry on the same side, of the pairs above it, if any such entry exists, takes its place.

When we enter a recursive step, a special "end of stack" marker E is placed on top of Π_2, and the three stacks are used as in the main algorithm. If the recursive step ends successfully, we first attempt to switch sections for each of the blocks with a nonempty section on Π_2, above the top most E. If we fail to expose E, then $C + B$ is nonplanar and we halt. Otherwise, all the blocks created during the recursion are joined to the one which includes v_j (the end vertex of the first path of B). The exposed E, on top of Π_2, is removed and we continue with the previous level of recursion. When we backtrack into a vertex v_i, all occurrences of v_i are removed from the top of Π_1 and Π_2, together with pairs of pointers of Π_3 which point to removed entries on both Π_1 and Π_2. (Technically, instead of pointing to an occurrence of v_i, we point to 0, and pairs (0, 0) are removed).

Theorem 8.1: The complexity of the path addition algorithm is $O(|V|)$.

Proof: As in the closing remarks of Section 8.1, we can assume $|E| = O(|V|)$. The DFS and the reordering of the adjacency lists have been shown to be $O(|V|)$. Each edge in the paths finding algorithm is used again, once in each direction. The total number of entries in the stacks Π_1 and Π_2 is bounded by the number of back edges ($|E| - |V| + 1$), and is therefore $O(|V|)$. After each section switching the number of blocks is reduced by one; thus the total work invested in section switchings is $O(|V|)$.

Q.E.D.

8.3   Computing an st-Numbering

In this section we shall define on st-numbering and describe a linear time algorithm to compute it. This numbering is necessary for the vertex addition algorithm, for testing planarity, of Lempel, Even and Cederbaum.

Given any edge $s - t$ of a nonseparable graph $G(V, E)$, a 1-1 function $g: V \rightarrow \{1, 2, \ldots, |V|\}$ is called an *st-numbering* if the following conditions are satisfied:

(1) $g(s) = 1$,
(2) $g(t) = |V| (=n)$,
(3) for every $v \in V - \{s, t\}$ there are adjacent vertices u and w such that $g(u) < g(v) < g(w)$.

Lempel, Even and Cederbaum showed that for every nonseparable graph and every edge $s - t$, there exists an st-numbering. The algorithm described here, following the work of Even and Tarjan [5], achieves this goal in linear time.

The algorithm starts with a DFS whose first vertex is t and its first edge is $t - s$. (i.e., $k(t) = 1$ and $k(s) = 2$). This DFS computes for each vertex v, its DFS number, $k(v)$, its father, $f(v)$, its lowpoint $L(v)$ and distinguishes tree edges from back edges. This information is used in the paths finding algorithm to be described next, which is different from the one used in the path addition algorithm.

Initially, s, t and the edge connecting them are marked "old" and all the other edges and vertices are marked "new". The path finding algorithm starts from a given vertex v and finds a path from it. This path may be directed from v or into v.

(1) If there is a "new" back edge $v \xrightarrow{e} w$ (in this case $k(w) < k(v)$) then do the following:

> Mark e "old".
> The path is $v \xrightarrow{e} w$.
> Halt.

(2) If there is a "new" tree edge $v \xrightarrow{e} w$ (in this case $k(w) > k(v)$) then do the following:

> Trace a path whose first edge is e and from there it follows a path which defined $L(w)$, i.e., it goes up the tree and ends with a back edge into a vertex u such that $k(u) = L(w)$. All vertices and edges on the path are marked "old". Halt.

(3) If there is a "new" back edge $w \xrightarrow{e} v$ (in this case $k(w) > k(v)$) then do the following:

> Start the path with e (going backwards on it) and continue backwards via tree edges until you encounter an "old" vertex. All vertices and edges on the path are marked "old". Halt.

(4) (All edges incident to v are "old"). The path produced is empty. Halt.

Lemma 8.9: If the path finding algorithm is always applied from an "old" vertex $v \neq t$ then all the ancestors of an "old" vertex are "old" too.

Proof: By induction on the number of applications of the path finding algorithm. Clearly, before the first application, the only ancestor of s is t

and it is old. Assuming the statement is true up to the present application, it is easy to see that if any of the four steps is applicable the statement continues to hold after its application.

Q.E.D.

Corollary 8.3: If G is nonseparable and under the condition of Lemma 8.9, each application of the path finding algorithm from an "old" vertex v produces a path, through "new" vertices and edges, to another "old" vertex, or in case all edges incident to v are "old", it returns the empty path.

Proof: The only case which requires a discussion is when case (2) of the path finding algorithm is applied. Since G is nonseparable, by Lemma 3.5, $L(w) < k(v)$. Thus, the path ends "below" v, in one of its ancestor. By Lemma 8.9, this ancestor is "old".

Q.E.D.

We are now ready to present the algorithm which produces an st-numbering. It uses a stack S which initially contains only t and s, s on top of t.

(1) $i \leftarrow 1$.
(2) Let v be the top vertex on S. Remove v from S. If $v = t$ then $g(t) \leftarrow i$ and halt.
(3) ($v \neq t$) Apply the path finding algorithm to v. If the path is empty then $g(v) \leftarrow i$, $i \leftarrow i + 1$ and go to Step (2).
(4) (The path is not empty) Let the path be $v - u_1 - u_2 - \cdots - u_l - w$. Put $u_l, u_{l-1}, \ldots, u_2, u_1, v$ on S in this order (v comes out on top) and go to Step (2).

Theorem 8.2: The algorithm above computes for every nonseparable graph $G(V, E)$ an st-numbering.

Proof: First we make a few observations about the algorithm:

(i) No vertex ever appears in two or more places on S at the same time.
(ii) Once a vertex v is placed on S, nothing under v receives a number until v does.
(iii) A vertex is permanently removed from S only after all its incident edges become "old".

Next, we want to show that each vertex v is placed on S before t is removed. Since t and s are placed on S initially, the statement needs to be proved for $v \neq s$, t only. Since G is nonseparable, there exists a simple path from s to v which does not pass through t (see Theorem 6.7 part (6)). Let this path be $s = u_1 - u_2 - \cdots - u_l = v$. Let m be the first index such that u_m is not placed on S. Since u_{m-1} is placed on S, t can be removed only after u_{m-1} (fact (ii)), and u_{m-1} is removed only after all its incident edges are "old" (fact (iii)). Thus, u_m must be placed on S before t is removed.

It remains to be shown that the algorithm computes an st-numbering. Since each vertex is placed on S, and eventually it is removed, each vertex v gets a number $g(v)$. Clearly, $g(s) = 1$, for it is the first to be removed. After each assignment i is incremented. Thus, $g(t) = |V|$. Every other vertex v is placed on S, for the first time, as an intermediate vertex on a path. Thus, there is an adjacent vertex stored below it, and an adjacent vertex stored above it. The one above it (by fact (ii)) gets a lower number and the one below it, a higher number.

$$Q.E.D.$$

It is easy to see that the whole algorithm is of time complexity $O(|E|)$: First, the DFS is $O(|E|)$. The total time spent on path finding is also $O(|E|)$ since no edge is used more than once. The total number of operations in the main algorithm is bounded also by $O(|E|)$ because the number of stack insertions is exactly $|E| + 1$.

8.4   The Vertex Addition Algorithm of Lempel, Even and Cederbaum

In this section we assume that $G(V, E)$ is a nonseparable graph whose vertices are st-numbered. From how on, we shall refer to the vertices by their st-number. Thus, $V = \{1, 2, \ldots, n\}$. Also, the edges are now directed from low to high.

A (graphical) *source* of a digraph is a vertex v such that $d_{in}(v) = 0$; a (graphical) *sink* is a vertex v such that $d_{out}(V) = 0$.* Clearly vertex 1 is a source of G and vertex n is a sink. Furthermore, due to the st-numbering, no other vertex is either a source or a sink.

Let $V_k = \{1, 2, \ldots, k\}$. $G_k(V_k, E_k)$ is the digraph induced by V_k, i.e., E_k consists of all the edges of G whose endpoints are both in V_k.

If G is planar, let $\hat{G}$ be a plane realization of G. It contains a plane realization of G_k. The following simple lemma reveals the reason for the st-numbering.

---

*Do not confuse with the source and sink of a network. The source of a network is not necessarily a (graphical) source, etc.

Lemma 8.10: If $\hat{G}_k$ is a plane realization of G_k contained in a plane digraph $\hat{G}$ then all the edges and vertices of $\hat{G} - \hat{G}_k$ are drawn in one face of $\hat{G}_k$.

Proof: Assume a face F of $\hat{G}_k$ contains vertices of $V - V_k$. Since all the vertices on F's window are lower than the vertices in F, the highest vertex in F must be a sink. Since G has only one sink, only one such face is possible.

$$\text{Q.E.D.}$$

Let B_k be the following digraph. G_k is a subgraph of B_k. In addition B_k contains all the edges of G which emanate from vertices of V_k and enter in G, vertices of $V - V_k$. These edges are called *virtual edges*, and the leaves they enter in B_k are called *virtual vertices*. These vertices are labeled as their counterparts in G, but they are kept separate; i.e., there may be several virtual vertices with the same label, each with exactly one entering edge. For example, consider the digraph shown in Fig. 8.1(a). B_3 of this digraph is shown in Fig. 8.1(b).

By Lemma 8.10, we can assume that if G is planar then there exist a plane realization of B_k in which all the virtual edges are drawn in the outside face. Furthermore, since $1 \overset{e}{\to} n$ is always an edge in G, if $k < n$ then vertex 1 is on the outside window and one of the virtual edges is e. In this case, we can draw B_k in the following form: Vertex 1 is drawn at the bottom level. All the virtual vertices appear on one horizontal line. The remaining vertices of G_k are drawn in such a way that vertices with higher names are

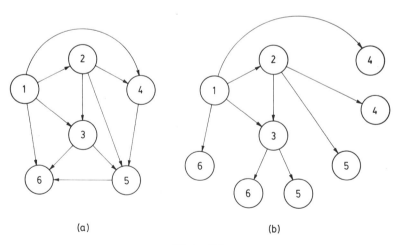

(a)   (b)

Figure 8.1

drawn higher. Such a realization is called a *bush form*. A bush form of B_3, of our example, is shown in Fig. 8.2.

In fact, Lemma 8.10 implies that if G is planar then there exists a bush form of B_k such that all the virtual vertices with labeled $k + 1$ appear next to each other on the horizontal line.

The algorithm proceeds by successively "drawing" $B_1, B_2, \ldots, B_{n-1}$ and G. If in the realization of B_k all the virtual vertices labeled $k + 1$ are next to each other, then it is easy to draw B_{k+1}: One joins all the virtual vertices labeled $k + 1$ into one vertex and "pulls" it down from the horizontal line. Now all the edges of G which emanate from $k + 1$ are added, and their other endpoints are labeled properly and placed in an arbitrary order on the horizontal line, in the space evacuated by the former virtual vertices labeled $k + 1$.

However, a difficulty arises. Indeed, the discussion up to now guarantees that if G is planar then there exists a sequence of bush forms, such that each one is "grown" from the previous one. But since we do not have a plane realization of G, we may put the virtual vertices, out of $k + 1$, in a "wrong" order. It is necessary to show that this does not matter; namely, by simple transformations it will be possible later to correct the "mistake".

Lemma 8.11: Assume v is a separation vertex of B_k. If $v > 1$ then exactly one component of B_k, with respect to v, contains vertices lower than v.

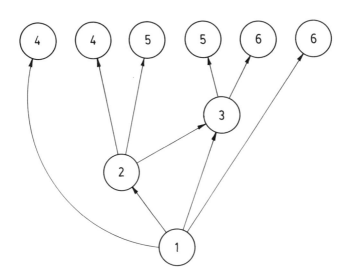

Figure 8.2

Note that here we ignore the direction of the edges, and the lemma is actually concerned with the undirected underlying graph of B_k.

Proof: The st-numbering implies that for every vertex u there exists a path from 1 to u such that all the vertices on the path are less than u. Thus, if $u < v$ then there is a path from 1 to u which does not pass through v. Therefore, 1 and u are in the same component.

$$\text{Q.E.D.}$$

Lemma 8.11 implies that a separation vertex v of B_k is the lowest vertex in each of the components, except the one which contains 1 (in case $v > 1$). Each of these components is a sub-bush; i.e. it has the same structure as a bush form, except that its lowest vertex is v rather than 1. These sub-bushes can be permuted around v in any of the $p!$ permutations, in case the number of these sub-bushes is p. In addition, each of the sub-bushes can be flipped over. These transformations maintain the bush form. It is our purpose to show that if $\hat{B}_k^1$ and $\hat{B}_k^2$ are bush forms of a planar B_k then through a sequence of permutations and flippings, one can change $\hat{B}_k^1$ into a $\hat{B}_k^3$ such that the virtual vertices of B_k appear in $\hat{B}_k^2$ and $\hat{B}_k^3$ in the same order.

For efficiency reasons, to be become clear to those readers who will study the implementation through $P\,Q$-trees, we assume that when a sub-bush is flipped, smaller sub-bushes of other separation vertices of the component, are not flipped by this action. For example, consider the bush form shown in Fig. 8.3(a). The bush form of Fig. 8.3(b) is achieved by permuting about 1, Fig. 8.3(c) by flipping about 1 and Fig. 8.3(d) by flipping about 2.

Lemma 8.12: Let H be a maximal nonseparable component of B_k and $y_1, y_2, \ldots, y_m$ be the vertices of H which are also endpoints of edges of $B_k - H$. In every bush form $\hat{B}_k$ all the y's are on the outside window of H and in the same order, except that the orientation may be reversed.

Proof: Since $\hat{B}_k$ is a bush form, all the y's are on the outside face of H. Assume there are two bush forms $\hat{B}_k^1$ and $\hat{B}_k^2$ in which the realizations of H are $\hat{H}^1$ and $\hat{H}^2$, respectively. If the y's do not appear in the same order on the outside windows of $\hat{H}^1$ and $\hat{H}^2$ then there are two y's, y_i and y_j which are next to each other in $\hat{H}^1$ but not in $\hat{H}^2$ (see Fig. 8.4). Therefore, in $\hat{H}^2$, there are two other y's, y_k and y_l which interpose between y_i and y_j on the two paths between them on the outside window of $\hat{H}^2$. However, from $\hat{H}^1$ we see that there are two paths, $P_1[y_i, y_j]$ and $P_2[y_k, y_l]$ which are completely disjoint. These two paths cannot exist simultaneously in $\hat{H}^2$. A contradiction.

$$\text{Q.E.D.}$$

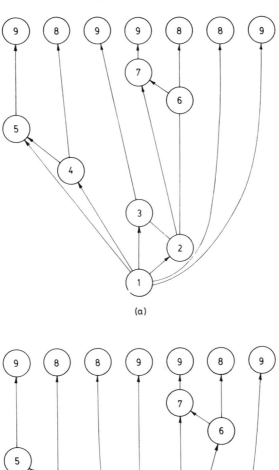

(a)

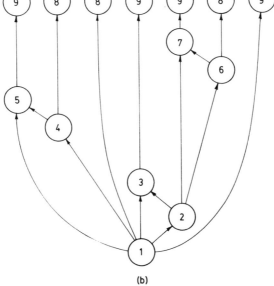

(b)

Figure 8.3 (a & b)

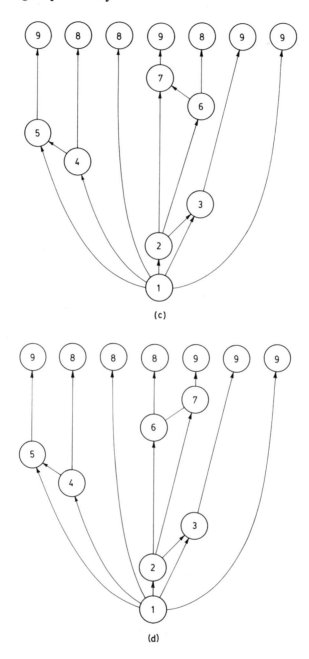

(c)

(d)

Figure 8.3 (c & d)

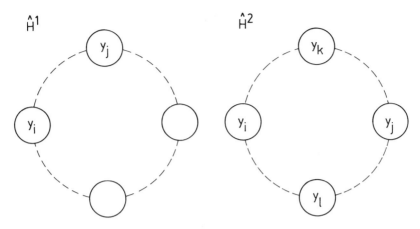

$\hat{H}^1$ $\hat{H}^2$

Figure 8.4

Theorem 8.3: If $\hat{B}_k{}^1$ and $\hat{B}_k{}^2$ are bush forms of the same B_k then there exists a sequence of permutations and flippings which transforms $\hat{B}_k{}^1$ into $\hat{B}_k{}^3$, such that in $\hat{B}_k{}^2$ and $\hat{B}_k{}^3$ the virtual vertices appear in the same order.

Proof: By induction on the size* of bush or sub-bush forms. Clearly, if each of the two (sub-)bushes consists of only one vertex and one virtual vertex, then the statement is trivial. Let v be the lowest vertex in the (sub-)bushes $\hat{B}^1$ and $\hat{B}^2$ of the same B. If v is a separation vertex, then the components of B appear as sub-bushes in $\hat{B}^1$ and $\hat{B}^2$. If they are not in the same order, by permuting them in $\hat{B}^1$, they can be put in the same order as in $\hat{B}^2$. By the inductive hypothesis, there is a sequence of permutations and flippings which will change each of the sub-bushs of $\hat{B}^1$ to have the order of its virtual vertices as in its counterpart in the $\hat{B}^2$, and therefore the theorem follows.

If v is not a separating vertex then let H be the maximal nonseparable component of B which contains v. In $\hat{B}^1(\hat{B}^2)$ there is a planar realization $\hat{H}^1(\hat{H}^2)$ of H. The vertices $y_1, y_2, \ldots, y_m$ of H, which are also endpoints of edges of $B - H$, by Lemma 8.12, must appear on the outside window of H in the same order, up to orientation. If the orientation of the y's in $\hat{H}^1$ is opposite to that of $\hat{H}^2$, flip the (sub-) bush $\hat{B}^1$ about v. Now, each of the y's is the lowest vertex of some sub-bush of B, and these sub-bushes appear in (the new) $\hat{B}^1$ and $\hat{B}^2$ in the same order. By the inductive hypothesis, each of these sub-bushes can be transformed by a sequence of permutations and

---

*The number of vertices.

flippings to have its virtual vertices in the same order as its counterpart in $\hat{B}^2$.

Q.E.D.

Corollary 8.4: If G is planar and $\hat{B}_k$ is a bush form of B_k then there exists a sequence of permutations and flippings which transforms $\hat{B}_k$ into a $\hat{B}_k{}'$ in which all the virtual vertices labeled $k + 1$ appear together on the horizontal line.

It remains to be shown how one decides which permutation or flipping to apply, how to represent the necessary information, without actually drawing bush forms, and how to do it all efficiently. Lempel, Even and Cederbaum described a method which uses a representation of the pertinent information by proper expressions. However, a better representation was suggested by Booth and Leuker. They invented a data structure, called PQ-trees, through which the algorithm can be run in linear time. PQ-trees were used to solve other problems of interest; see [6].

I shall not describe PQ-trees in detail. The description in [6] is long (30 pages) although not hard to follow. It involves ideas of data structure manipulation, but almost no graph theory. The following is a brief and incomplete description.

A PQ-tree is a directed ordered tree, with three types of vertices: P-vertices, Q-vertices and leaves. Each P-vertex, or Q-vertex, has at least one son. The sons of a P-vertex, which in our application represents a separating vertex v of B_k, may be permuted into any new order. Each of the sons, and its subtree, represents a sub-bush. A Q-vertex represents a maximal non-separable component and its sons which represent the y's, may not be permuted, but their order can be reversed. The leaves represent the virtual vertices.

The attempt to gather all the leaves labeled $k + 1$ into an unbroken run, is done from sons to fathers, starting with the leaves labeled $k + 1$. Through a technique of template matching, vertices are modified while the $k + 1$ labeled leaves are bunched together. Only the smallest subtree which contains all the $k + 1$ labeled leaves is scanned. All these leaves, if successfully gathered, are merged into one P-vertex and its sons represent the virtual edges out of $k + 1$. This procedure is repeated until $k + 1 = n$.

PROBLEMS

8.1 Demonstrate the path addition algorithm on the Peterson graph (see problem 7.3). Show the data for all the steps: The DFS for numbering the vertices, defining the tree and computing $L1$ and $L2$. The ϕ function on the edges. The sorting of the adjacency lists. Use the path finding algorithm in the new DFS, to decompose the graph into C and a sequence of paths. Use Π_1, Π_2, Π_3 and end of stack markers to carry out all the recursive steps up to planarity decision.

8.2 Repeat the path addition planarity test, as in Problem 8.1, for the graph given below.

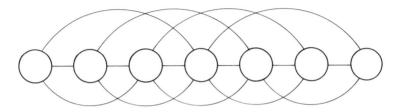

8.3 Demonstrate the vertex addition planarity test on the Peterson graph. Show the steps for the DFS, the st-numbering and the sequence of bush forms.

8.4 Repeat the vertex addition planarity test for the graph of Problem 8.2.

8.5 Show that if a graph is nonplanar then a subgraph homeomorphic to one of the Kuratowski's graphs can be found in $O(|V|^2)$. (Hints: Only $O(|V|)$ edges need to be considered. Delete edges if their deletion does not make the graph planar. What is left?)

REFERENCES

[1] Auslander, L., and Parter, S. V., "On Imbedding Graphs in the Plane," J. Math. and Mech., Vol. 10, No. 3, May 1961, pp. 517-523.

[2] Goldstein, A. J., "An Efficient and Constructive Algorithm for Testing Whether a Graph Can be Embedded in a Plane," Graph and Combinatorics Conf., Contract No. NONR 1858-(21), Office of Naval Research Logistics Proj., Dept. of Math., Princeton Univ., May 16-18, 1963, 2 pp.

[3] Hopcroft, J., and Tarjan, R., "Efficient Planarity Testing," JACM, Vol. 21, No. 4, Oct. 1974, pp. 549-568.

[4] Lempel, A., Even, S., and Cederbaum, I., "An Algorithm for Planarity Testing of Graphs," *Theory of Graphs, International Symposium, Rome,* July, 1966. P. Rosenstiehl, Ed., Gordon and Breach, N.Y. 1967, pp. 215-232.

[5] Even, S., and Tarjan, R. E., "Computing an st-numbering," Th. Comp. Sci., Vol. 2, 1976, pp. 339-344.

[6] Booth, K. S., and Lueker, G. S., "Testing for the Consecutive Ones Property, Interval Graphs, and Graph Planarity Using PQ-tree Algorithms," J. of Comp. and Sys. Sciences, Vol. 13, 1976, pp. 335-379.

Chapter 9

THE THEORY OF NP-COMPLETENESS

9.1 Introduction

For many years many researchers have been trying to find efficient algorithms for solving various combinatorial problems, with only partial success. In the previous chapters several of the achievements were described, but many problems arising in areas such as computer science, operations research, electrical engineering, number theory and other branches of discrete mathematics have defied solution in spite of the massive attempt to solve them. Some of these problems are: The simplification of Boolean functions, scheduling problems, the traveling salesman problem, certain flow problems, covering problems, placement of components problems, minimum feedback problems, prime factorization of integers, minimum coloration of graphs, winning strategies for combinatorial games.

In this chapter we shall introduce a class of problems, which includes hundreds of problems which have been attempted individually, and no efficient algorithm has been found to solve any of them. Furthermore, we shall show that a solution of any one member of this class, will imply a solution for all. This is no direct proof that members of this class are hard to solve, but it provides a strong circumstantial evidence that such a solution is unlikely to exist.

Since all the problems we consider are solvable, in the sense that there is an algorithm for their solution (in finite time), we need a criterion for deciding whether an algorithm is efficient. To this end, let us discuss what we mean by the *input length*. For every problem (for example, deciding whether a given integer is a prime) we seek an algorithm such that for every instance of this problem (say, 5127) the algorithm will answer the question ("is it a prime"?) correctly. The length of the data describing the instance is called the input length (in our illustration it is 4 decimal digits). This length depends on the format we have chosen to represent the data; i.e. we can use decimal notation for integers, or binary notation (13 bits for our illustration) or any other well defined notation; for graphs, we can use adjacency matrices, or incidence lists, etc.

Following Edmonds [1], we say that an algorithm is efficient if there exists a polynomial $p(n)$ such that an instance whose input length is n takes at most $p(n)$ elementary computational steps to solve. That is, we accept an algorithm as efficient only if it is of polynomial (time) complexity. This is clearly a very crude criterion, since it says nothing about the degree or coefficients of the polynomial. However, we accept it as a first approximation for the complexity of a problem for the following reasons:

(1) All the problems which are considered efficiently solved, in the literature, have known polynomial algorithms for their solutions. For example, all the algorithms of the previous chapters of this book are polynomial.

(2) None of the hard problems, mentioned above, is known to have a polynomial algorithm.

(3) To see why polynomial is better than, say, exponential, consider the following situation. Assume we have two algorithms for a solution of a certain problem. Algorithm A is of complexity n^2 and algorithm B is of complexity 2^n. Assume, we have a bound of say, one hour computation time, and we consider an instance manageable with respect to a certain algorithm if it can be solved by this algorithm within one hour. Let n_0 be the longest instance (n_0 is its input length) which can be solved by algorithm B, using a given computer C. Thus, the number of steps C can perform in one hour is 2^{n_0}. Now, if we buy a new computer C', say 10 times faster, the largest instance n we can handle satisfies

$$2^n = 10 \cdot 2^{n_0}$$

or

$$n = n_0 + \log_2 10.$$

This means, that instead of, say, factoring integers of n_0 binary digits we can now factor integers of about $n_0 + 4$ digits. This is not a very dramatic improvement.

However, if n_0 is the largest instance we could handle, by C, using Algorithm A, then now we can handle, by C', instances of length up to n, where

$$n^2 = 10 \cdot n_0^2,$$

and

$$n = \sqrt{10} \cdot n_0.$$

This means that we would be able to factor integers with more than 3 times the number of digits as we could before. This is much more appealing. (Unfortunately, we do not know of an n^2 algorithm for factoring integers.)

Having a problem in mind, we have to decide on the format of its input. Only after this is done, the question of whether there is a polynomial algorithm is mathematically defined. We do not mind exactly what format one chooses for our problem, up to certain limits. We consider two formats equally acceptable if each can be translated to the other in polynomial time. If in one format the input length of an instance is n and $q(n)$ is a polynomial bound on the time to translate it (by the translation algorithm) to the second format, then clearly the input length in the second format of the same instance is bounded by $q(n)$. Now, if we have a polynomial time algorithm to solve the problem in the second format, where the complexity is bounded by $p(n)$, then there exists a polynomial time algorithm for the first format too: First we apply the translation algorithm and to its output we apply the polynomial algorithm for the second format. The combined algorithm is of complexity

$$q(n) + p(q(n)),$$

which is still polynomial.

However, if no such polynomial translation algorithm exists, then we do not consider the two formats equally acceptable. For example, the binary representation of integers and the unary representation (13 is represented by 1111111111111) are not equally acceptable, since the length of the second is exponential, and not polynomial, in terms of the first. It is easy to describe a polynomial algorithm for factoring integers given in unary ($O(n^2)$, where n in this format is not only the input length but also the value of the integer; how can this be done?). This does not imply a polynomial algorithm for the binary representation. In this case we consider the binary representation as acceptable, and the unary representation—unacceptable.

A problem is called a decision problem if the answer is 'yes' or 'no'; for example, "is a given integer a prime?" is a decision problem. However, in many problems the answer required may be more involved. For example, "find the prime factorization of a given integer", or "what is the maximum flow possible in a given network?", require longer answers. For reasons to become clear later, we prefer to represent each of our problems by a closely related decision problem. For example, the problem of prime factorization is replaced by the following: "Given an integer n and two more integers a and b, is there a prime factor f of n such that $a \leq f \leq b$?" The problem of finding the maximum flow of a network can be replaced by: "Given a network N and an integer k, is there a legal flow in N whose total value is at least k?" Clearly, having a polynomial algorithm which answers the more involved problem implies the existence of a polynomial algorithm to answer the corresponding decision problem. [For example, if we have a polynomial algorithm for prime factorization, we can examine each factor (how many can an integer have if its binary representation is of length n?) and check if any of them satisfies the condition.] The converse is less trivial, but still holds; i.e. if we have a polynomial algorithm to solve the decision problem we can solve the original problem in polynomial time too. In the case of factoring an integer N (its input length is $n = \lceil \log_2 (N + 1) \rceil$), we first ask whether there is a prime factor inbetween 2 and $\lfloor N/2 \rfloor$ (actually $\sqrt{N}$ is sufficient, but this makes no difference in our discussion). If the answer is positive we divide the region into 2, etc. continuing in a binary search manner. In $O(n)$ repeated applications of the decision problem we find one factor. We divide N by the factor found and repeat the process for the quotient. Finding all factors requires at most $O(n^2)$ applications of the decision problem. Thus, if $O(p(n))$ is the complexity of the decision problem, then the factorization is $O(n^2 p(n))$. In most cases, the corresponding algorithm is even simpler than in this demonstration.

9.2 The NP Class of Decision Problems

We need a more precise model for an algorithm. This is provided by the Turing machine model. The readers are assumed to be acquainted with Turing machines and Church's Thesis (see any standard textbook on theory of computation, such as [2] or [3]). Here we shall define a Turing machine, since we need its exact formulation in what follows, but we assume the reader is familiar with the thesis that a problem can be solved by an algorithm if and only if it can be solved by some Turing machine which halts for every input.

A Turing machine M is defined by an eight-tuple $(S, \Gamma, \Sigma, f, b, s_0, s_Y, s_N)$ where

S is a finite set whose elements are called *states*; s_0, s_Y, s_N are three states called the *initial state*, the *'yes' state* and the *'no' state*, respectively.

Γ is a finite set whose elements are called *tape symbols*; Σ is a proper subset of Γ whose elements are called *input symbols*; b is an element of Γ-Σ, called the *blank symbol*.

f is a function $(S - \{s_Y, s_N\}) \times \Gamma \rightarrow S \times \Gamma \times \{1, -1\}$, called the *transition* function.

A Turing machine has an infinite tape, divided into cells $\cdots c(-2)$, $c(-1)$, $c(0)$, $c(1)$, $c(2) \cdots$. The machine has a read-write head. Each unit of time t, the head is located at one of the tape cells; the fact that at time t the head is in $c(i)$ is recorded by assigning the *head location function* h the value i ($h(t) = i$). When the computation starts, $t = 0$. The head is initially in $c(1)$; i.e. $h(0) = 1$.

Each cell contains one tape symbol at a time. The tape symbol of $c(i)$ at time t is denoted by $\gamma(i, t)$. The *input data* consists of input symbols x_1, x_2, $\ldots$, x_n and is initially written on the tape in cells $c(1)$, $c(2)$, $\ldots$, $c(n)$ respectively; i.e. $\gamma(1, 0) = x_1$, $\gamma(2, 0) = x_2$, $\ldots$, $\gamma(n, 0) = x_n$. For all other cells $c(i)$, $\gamma(i, 0) = b$.

The state at time t is denoted by $s(t)$. Assume $s(t) \in S - \{s_Y, s_N\}$, and $f(s(t), \gamma(h(t), t)) = (p, q, d)$. In this case, the next state $s(t + 1) = p$, the symbol in $c(h(t))$ becomes q ($\gamma(h(t), t + 1) = q$) and the head location moves one cell to the right (left) if $d = 1(-1)$; i.e. $h(t + 1) = h(t) + d$. All other cells $c(i)$, $i \neq h(t)$, retain their symbol; i.e. $\gamma(i, t + 1) = \gamma(i, t)$. If $s(t) \in \{s_Y, s_N\}$ the machine halts.

We assume that M has the property that for every input data it eventually halts.

A Turing machine M is said to solve a decision problem P, if for every instance of this problem, described according to the conventional format by $x_1, x_2, \ldots, x_n$ (n is the input length), the answer is 'yes' if and only if M, when applied to $x_1, x_2, \ldots, x_n$ as its input data, halts in state s_Y.

A Turing machine seems to be a very slow device for computation. Typically, a lot of time is "wasted" on moving the head from one location to another, step by step. This is in contrast to the random access machine model which we usually use for our computations. Yet, it can be shown that if an algorithm solves a problem on a random access machine in polynominal time then a corresponding Turing machine will solve this problem in polynomial time too. (The new polynomial may be of higher degree, but this does not concern us.)

There is a rich set of decision problems for which we do not know of a polynomial algorithm (or Turing machine) for their solution, but a positive answer can be verified in polynomial time if some additional information is given. For example, if we are given an integer N for which the question: "Is it true that N is not a prime?" has to be answered, and if a factor F of N is given, it is easy to verify, in polynomial time, that N is not a prime by simply verifying that F is indeed a factor. Another example is the following. Assume we are given an undirected graph $G(V, E)$ and want to answer the question: "Does G have a Hamilton circuit?" (see Problem 1.2). If in addition a Hamilton circuit is specified, it is easy to verify in polynomial time that indeed the given circuit is Hamiltonian. To capture this concept more formally, we use nondeterministic Turing machines.

Assume that before our machine M starts its computation, some information is placed in cells $c(-1)$, $c(-2)$, ... on the tape; i.e. these cells are not restricted to contain initially blanks only. The sequence of symbols

$$\gamma(-1, 0), \gamma(-2, 0), \ldots = g_1, g_2, \ldots$$

is called the *guess*. Clearly, if the time alloted to M is bounded by some polynomial $p(n)$, then there is no point in placing a guess longer than $p(n) - 1$, since $c(-p(n))$ cannot be reached at all during the computation.

We say that a decision problem D can be solved by a *nondeterministic Turing machine M* in polynomial time if there exists a polynomial $p(n)$ such that if the answer to an instance I of input length n is 'yes' then there exists a guess for which M will halt in $p(n)$ units of time, in state s_Y; and if the answer is 'no' then for every guess either M will not halt at all, or if it halts, it will be in state s_N.

The reason for calling such a "computation" nondeterministic is that we do not claim to have a (deterministic) method to make the guess. We merely say that one exists. The only obvious way to convert a nondeterministic machine into an algorithm (or a deterministic Turing machine) is to try M on all $|\Gamma|^{p(n-1)}$ possible guesses and see if any of them will give a positive answer. Unfortunately, the time necessary is exponential ($p(n) \cdot |\Gamma|^{p(n)-1}$).

The reader should realize that there are polynomial nondeterministic Turing machines to solve the problems which we mentioned as "easy" to verify. In the case of "Is it true that N is not a prime?" the guess is simply a specification of a factor F. The machine divides N by F and verifies that indeed the remainder is zero. It then enters s_Y and halts. If the remainder is not zero, it enters s_N and halt. (This does not mean that N is a prime. It merely means that F is not one of its factors.)

In the case of the Hamilton circuit problem, the guess is a specification of a circuit. The machine checks whether this is indeed a Hamilton circuit, i.e. it is a circuit, it is simple and passes through all the vertices. If all the tests end positively, it halts in state s_Y. If the guess does not have the agreed format, or if it is not a circuit, or if it is not simple or if it does not include all the vertices, then M halts in state s_N, (Again, if it halts in s_N, this does not mean that G has no Hamilton circuit. All it says is that the given guess is a failure.)

Thus, a polynomial nondeterministic Turing machine can be an efficient device to get a positive answer, if we are lucky with the guess, but is a very ineffective device for getting a negative answer to our decision problem.

Let P be the set of decision problems which can be solved by some polynomial Turing machine (a deterministic Turing machine for which a polynomial time bound $p(n)$ exists).

Let NP be the set of decision problems which can be solved by some nondeterministic Turing machine with a polynomial time bound.

Clearly, $P \subseteq NP$, since we can use the machine M which solves the problem deterministically with the blank guess.

The question of whether $P = NP$ has not been answered yet. It is generally believed that $P \neq NP$; that is, there are problems in NP which cannot be solved (deterministically) in polynomial time. However, we shall not rely on this assumption. Instead, one can view what follows in this chapter and the next, as circumstantial evidence that $P \neq NP$.

9.3   NP-Complete Problems and Cook's Theorem

Let D_1 and D_2 be decision problems. We say that there exists a *polynomial reduction* of D_1 to D_2 ($D_1 \propto D_2$) if there exists a function $f(I_1)$ from the set of inputs of D_1 to the set of inputs of D_2, such that the answer to I_1 is 'yes', with respect to D_1, if and only if the answer to $f(I_1)$ is 'yes' with respect to D_2, and there exists a polynomially bounded algorithm to compute $f(I_1)$. Such an algorithm is realizable by a Turing machine which computes $f(I_1)$; i.e. unlike the machines of the previous section which only produce a 'yes' or 'no' answer, this machine prints on the tape the value of $f(I_1)$ before it halts.

If $D_1 \propto D_2$ and D_2 can be answered by a polynomially bounded algorithm A_2 then D_1 is also solvable by a polynomially bounded algorithm: Given an input I_1 for D_1, use first the polynomially bounded algorithm to produce $f(I_1)$, and assume this computation is bounded by the polynomial $q(n)$, where n is the length of I_1. Now, use A_2 to answer $f(I_1)$ with respect to D_2.

Let $p(m)$ the polynomial bounding the computation time of A_2, where m is the length of $I_2 = f(I_1)$. Since $m \leq q(n)$, the total computation time to answer D_1 is bounded by $q(n) + p(q(n))$, which is clearly a polynomial.

Following Karp's approach [4], let us define a decision problem D to be *NP-Complete* (NPC) if $D \in NP$ and for every $D' \in NP$, $D' \propto D$.

Clearly, if D is NPC and if there is a polynomial algorithm to solve it ($D \in P$) then every $D' \in NP$ is also in P, or $P = NP$.

Note also that the relation $\propto$ is transitive; i.e. if $D_1 \propto D_2$ and $D_2 \propto D_3$ then $D_1 \propto D_3$. If $f_1(I_1)$ is a reduction function from D_1 to D_2 and $f_2(I_2)$ is a reduction function from D_2 to D_3 then $f_2(f_1(I_1))$ is a reduction function from D_1 to D_3, and if both f_1 and f_2 are polynomially computable, then so is $f_2(f_1)$.

It follows that if D is NPC, $D' \in NP$ and $D \propto D'$, then D' is also NPC.

Before we continue our study of the class of NPC problems it is worth noting that there are two other definitions of NPC in the literature.

According to Cook's approach [5], a problem $D \in NP$ is NPC if for every problem $D' \in NP$ there exists a polynomial algorithm in which questions of the type: "What is the answer to I with respect to D?" can be asked, and the answer be used. He calls such a question answering device an *oracle*, and its use is like a subroutine. Clearly, if there is a polynomial algorithm for some D which is NPC according to this definition, then $P = NP$. For if each of the calls to the oracle takes a polynomial time to answer, and there can be only a polynomial number of such calls, then the whole process is polynomially bounded.

Aho, Hopcroft and Ullman [6] use a more "general" definition yet. They call a problem D NPC if $D \in NP$ and the existence of a polynomial algorithm for it implies a polynomial algorithm for every $D' \in NP$. Clearly

$$NPC_{KARP} \subseteq NPC_{COOK} \subseteq NPC_{AHU}.$$

None of these inclusions is known to be strict, but equality has not been demonstrated either.

In certain studies the differences between these different definitions may be crucial, and the reader is advised to be aware of the differences. However, these differences are irrelevant in what follows in this book. For definiteness, we shall stick to the Karp approach.

Our aim is to prove that several combinatorial decision problems, which are of wide interest, are NPC. Typically, it is easy to show that they belong to NP, and as noted before, once we know one NPC problem D, a demonstration of a polynomial reduction of D to the new problem will establish its NP-Completeness. We need one NPC problem to start this approach off.

The problem which was chosen by Cook [4] to be the first established NPC problem is the satisfiability of conjunctive normal forms (SAT)*.

SAT is defined as follows. There is a finite set $X = \{x_1, x_2, \ldots, x_n\}$ of *variables*. A *literal* is either a variable x_i or its *complement* $\bar{x}_i$. Thus, the set of literals is $L = \{x_1, x_2, \ldots, x_n, \bar{x}_1, \bar{x}_2, \ldots, \bar{x}_n\}$. A *clause* C is a subset of L. We are given a set of clauses $C_1, C_2, \ldots, C_m$. The question is whether the set of variables can be assigned 'true' (T) or 'false' (F) values, so that each clause contains at least one literal with a T value. Clearly, if a literal is assigned a $T(F)$ value, its value is the same in all the clauses in which it appears. Also, if $x_i = T$ then $\bar{x}_i = F$, and if $x_i = F$ then $\bar{x}_i = T$. The requirement that the assignment of truth values satisfies these last two conditions is called the *consistency condition*. The requirement that each clause contains at least one literal assigned T is called the *satisfiability condition*.

A concise statement of SAT is, therefore, the following:

Input: A set of clauses.

Question: Is there a consistent assignment of truth values to the literals in these clauses, which meets the satisfiability condition?

Theorem 9.1: SAT is NPC.

Proof: First, it is easy to see that SAT ϵ NP. For if the set of clauses is satisfiable, all we need to do is to use a satisfying truth assignment as the guess and verify that indeed this assignment is consistent and satisfying. A Turing machine which performs this task can be built; this is a cumbersome but straightforward task.

The more demanding part of the proof is to show that every NP problem is polynomially reducible to SAT. Although this proof is long, its idea is ingeniously simple.

By definition, for every decision problem D in NP, there exists a polynomially bounded nondeterministic Turing machine M which solves it. We display a polynomial reduction such that when M, its bounding polynomial

---

*In fact Cook used the tautology problem of disjunctive normal forms. We use SAT, which by De Morgan's law is the complement of the tautology problem, since the latter is not known to be in NP.

$p(n)$ and the input I (for D) are given, it constructs an instance $f(I)$ of SAT, such that $f(I)$ is satisfiable if and only if M "accepts" I; i.e. there exists a guess with which M will verify that the answer to I, with respect to D, is 'yes' and the computation time be at most $p(n)$.

The idea is that $f(I)$ simulates the operation of M on the instance I. There will be eight sets of clauses. The satisfiability of each set S_i assures that a certain condition holds. The conditions are:

(1) Initially, I is specified by the contents of cells $c(1)$, $c(2)$, ..., $c(n)$; the cells $c(0)$ and $c(n + 1)$, $c(n + 2)$, ..., $c(p(n))$ all contain blanks.

(2) The initial state is s_0 and the head is in $c(1)$ ($h(0) = 1$).

(3) For every $0 \le t \le p(n)$, the machine is exactly in one state.

(4) For every unit of time t, $0 \le t \le p(n)$, each cell $c(i)$, $-p(n) + 1 \le i \le p(n) + 1$, contains exactly one tape symbol.

(5) For every $0 \le t \le p(n)$, the head is exactly on one cell $c(i)$, $-p(n) + 1 \le i \le p(n) + 1$.

(6) The contents of each cell can change from time t to time $t + 1$ only if at time t the head is on this cell.

(7) If $s(t) \in S - \{s_Y, s_N\}$ then the next state ($s(t + 1)$), the next tape symbol in the cell under the head ($\gamma(h(t), t + 1)$) and the next location of the head ($h(t + 1)$) are according to f. If $s(t) \in \{s_Y, s_N\}$ then $s(t + 1) = s(t)$. (We assume that the state does not change after M halts.)

(8) At time $p(n)$ the state is s_Y.

To simplify our notation let $S = \{s_0, s_1, s_2, \ldots, s_q\}$, where s_1 is the 'yes' state (s_Y) and s_2 is the 'no' state (s_N). We shall use the following conventions. Index i is used for cells; $-p(n) + 1 \le i \le p(n) + 1$. Index t is used for time; $0 \le t \le p(n)$. Index j is used for tape symbols; $0 \le j \le g$, where $\Gamma = \{\gamma_0, \gamma_1, \ldots, \gamma_g\}$ and γ_0 is the blank symbol (b). Index k is used for states; $0 \le k \le q$. The set of variables of the satisfiability problem is:

$$\{G(i, t, j)\} \cup \{H(i, t)\} \cup \{S(t, k)\}.$$

Since the number of tape symbols is fixed (for the given M, which does not change with I) the number of G variables is $0((p(n))^2)$. The number of H variables is also $0((p(n))^2)$ and the number of S variables is $0(p(n))$. The interpretation is as follows:

$$G(i, t, j) = T \text{ if and only if } \gamma(i, t) = \gamma_j;$$

i.e. the tape symbol in $c(i)$ at time t is γ_j.

$h(i, t) = T$ if and only if the head is on $c(i)$ at time t.

$S(t, k) = T$ if and only if the state at time t is s_k.

Assume that $I = x_1x_2x_3 \ldots x_n$. The set S_1 is given by:

$$S_1 = \{\{G(i, 0, j)\} \mid 1 \leq i \leq n \text{ and } x_i = \gamma_j\} \cup$$
$$\{\{G(i, 0, 0)\} \mid n < i \leq p(n)\} \cup \{\{G(0,0,0)\}\}.$$

It is easy to see that condition (1) holds if and only if all the clauses of S_1 are satisfied. The fact that the values of $G(i, 0, j)$ are not specified for $i < 0$ corresponds to the fact that one may use any guess one wants.

$$S_2 = \{\{S(0, 0)\}, \{H(1, 0)\}\}.$$

$$S_3 = \bigcup_t [\{\{S(t, 0), S(t, 1), \ldots, S(t, q)\}\} \cup \bigcup_{k_1 < k_2} \{\{\overline{S(t, k_1)}, \overline{S(t, k_2)}\}\}].$$

For every t S_3 contains one clause which guarantees that M is in a state and for each pair of states there is a clause which assures that M is never in two states at one time

$$S_4 = \bigcup_t \bigcup_i [\{\{G(i, t, 0), G(i, t, 1), \ldots, G(i, t, g)\}\}$$
$$\cup \bigcup_{j_1 < j_2} \{\{\overline{G(i, t, j_1)}, \overline{G(i, t, j_2)}\}\}].$$

$$S_5 = \bigcup_t [\{\{H(-p(n) + 1, t), H(-p(n) + 2, t), \ldots, H(p(n) + 1, t)\}\}$$
$$\cup \bigcup_{i_1 < i_2} \{\{\overline{H(i_1, t)}, \overline{H(i_2, t)}\}\}].$$

$$S_6 = \bigcup_i \bigcup_{0 \leq t < p(n)} \bigcup_j \{\{H(i, t,), G(i, t, j), \overline{G(i, t + 1, j)}\}\}.$$

For every i and (relevant) t there are g clauses which together require that either the head is on cell $c(i)$ at time t or the symbol $\gamma(i, t)$ is the same as $\gamma(i, t + 1)$; for if at time $t + 1$ it is j and at time t it is not, then no literal in the clause for i, t, j is 'true'.

$$S_7 = \bigcup_i \bigcup_{0 \leq t < p(n)} \bigcup_{k \neq 1, 2} \bigcup_j \{\{\overline{S(t, k)}, \overline{H(i, t)}, \overline{G(i, t, j)}, S(t + 1, k')\},$$
$$\{\overline{S(t, k)}, \overline{H(i, t)}. \overline{G(i, t, j)}, G(i, t + 1, j')\},$$
$$\{\overline{S(t, k)}, \overline{H(i, t)}, \overline{G(i, t, j)}, H(i + d, t + 1)\} \mid f(s_k, \gamma_j) = (s_{k'}, \gamma_{j'}, d)\}$$
$$\cup \bigcup_{t < p(n)} \{\{\overline{S(t, 1)}, S(t + 1, 1)\}, \{\overline{S(t, 2)}, S(t + 1, 2)\}\}.$$

The clauses of the last line imply that if $s(t) = s_1$ or s_2 (the 'yes' and 'no' states) then $s(t + 1) = s(t)$. The other three lines imply that if $s(t) = s_k$, $h(t) = c(i)$ and $\gamma(c(i), t) = \gamma_j$ then $s(t + 1)$, $\gamma(c(i), t + 1)$ and $h(t + 1)$ are according to the transition function f of the Turing machine M.

$$S_8 = \{\{S(p(n), 1)\}\}.$$

It follows that the set of clauses $\bigcup_{i=1}^{8} S_i$ is satisfiable if and only if there is a guess for which M accepts the input I, by reaching by $t = p(n)$ the 'yes' state. Thus, every NP decision problem is polynomially reducible to SAT.

Q.E.D.

The 3SAT problem is a subproblem of SAT, and yet it is also NPC. This is a useful result since for many NP-Completeness proofs it is easier to demonstrate a polynomial reduction from 3SAT, instead of using SAT directly. The definition of 3SAT is as follows:

Input: A set of clauses, each consisting of exactly three literals.

Question: Is there a consistent assignment of truth values to the literals in the clauses, which meets the satisfiability condition?

Theorem 9.2: 3SAT is NPC.

Proof. The proof that 3SAT ϵ NP is the same as for SAT. Next we show that SAT $\propto$ 3SAT.

Let us demonstrate the idea of the reduction on a simple example. Let S_1 be a set of clauses one of which is $\{a, b, c, d\}$; i.e. $S_1 = \{\{a, b, c, d\}\}$ $\cup S'$. Let x be a variable which does not appear in S_1. Define $S_2 = \{\{a, b, x\}, \{\bar{x}, c, d\}\} \cup S'$. Let us show that S_1 is satisfiable if and only if S_2 is. If a truth assignment in S_1 satisfies it, then the same assignment in S_2 satisfies S' and at least one of the first two clauses of S_2, even if x and $\bar{x}$ are ignored. Now x can be assigned a truth value which satisfies the remaining unsatisfied clause, if any. The resulting assigment satisfies S_2.

If a truth assignment satisfies S_2 then clearly it satisfies S'. Also, the variable x can cause the satisfiability of only one of the first two clauses. Thus one of the literals a, b, c, d is assigned a 'true' value. It follows that the same assignment, ignoring x, satisfies S_1 too.

This example shows how a clause of length four can be replaced by two clauses of length three. By introducing $l - 3$ new variables, a clause of of length $l > 3$ can be replaced by $l - 2$ clauses of length three, without changing the satisfiability of the set of clauses. This can be done as follows: The clause $\{a_1, a_2, \ldots, a_l\}$ is replaced by

$$\{a_1, a_2, x_1\}, \{\overline{x_1}, a_3, x_2\}, \ldots, \{\overline{x_i}, a_{i+2}, x_{i+1}\}, \ldots, \}\overline{x}_{l-3}, a_{l-1}, a_l\}.$$

If a truth assignment for $a_1, a_2, \ldots, a_l$ contains at least one 'true' literal then the long clause $\{a_1, a_2, \ldots, a_l\}$ is satisfied. Furthermore, the x variables can be assigned to satisfy all the clauses of length three which are not satisfied by one of the a-s: If a_1 or a_2 is T, assign $x_1 = x_2 = \ldots = x_{l-3} = F$. If a_{l-1} or a_l is T, assign $x_1 = x_2 = \cdots = x_{l-3} = T$. If some $a_k = T$, $2 < k < l - 1$, assign $x_1 = x_2 = \cdots = x_{k-2} = T$ and $x_{k-1} = x_k = \cdots x_{l-3} = F$. At least one of these cases applies. However, if a truth assignment for $a_1, a_2, \ldots, a_l$ makes them all 'false' then the long clause is not satisfied and no choice of values for the x variables will satisfy all the $l - 2$ short clauses.

Note that this replacement of long clauses by short ones can be done in time which is bounded polynomially with the length of the input which describes the original set of clauses. The number of clauses in the transformed set is bounded by the number of literals, counting their repetitions, in the original set, and each of the clauses in the transformed set contains at most three literals.

Q.E.D.

9.4   Three Combinatorial Problems which are NPC

Although this book deals mainly with graph theoretic problems, we shall discuss in this section three nongraphic combinatorial problems, and prove their *NP*-Completeness. These problems are important and well known, and their established *NP*-Completeness will help in the analysis of some graphic problems in the next chapter.

The first problem is the *three dimensional matching* (3DM). In Section 6.4 we discussed the problem of maximum matching in bipartite graphs, and saw that it can be solved in polynomial time; this problem is also called the two dimensional matching problem (2DM). The 3DM is defined as follows:

Input: Let W, X, and Y be three sets all of the same cardinality, p, and $M \subseteq W \times X \times Y$.

Question: Is there a subset M' of M, of cardinality p, such that no two triples in M' agree in any of their components?

For example, let $W = X = Y = \{0, 1\}$ and $M = \{(0, 0, 0), (0, 0, 1),$ $(0, 1, 0), (1, 0, 0)\}$. For this instance of the input the answer to the 3DM is 'no', since two triples of M agree in at least one component. However, if we change M by adding to it the triple $(1, 0, 1)$, then the answer is 'yes', since $M' = \{(0, 1, 0), (1, 0, 1)\}$ satisfies the condition.

Theorem 9.3: 3DM is NPC.

Proof. Clearly 3DM is in *NP*. Let us show that 3SAT $\propto$ 3DM, and since 3SAT is already known to be NPC, this will prove that 3DM is NPC. The proof follows Garey and Johnson [7]*.

Let $x_1, x_2, \ldots, x_n$ be the variables which appear in the set C of clauses $C_1, C_2, \ldots, C_m$, which is the input I to the 3SAT problem.

We construct W, X, Y and M of $f(I)$, the input to 3DM, in the following general manner. There will be a set of triples, AC, representing the truth assignment consistency of the variables. Here the consistency means that in all its appearances in C a variable gets the same truth value, and all the appearances of its complement get the complementary truth value. There will be a set of triples, SC, representing the satisfiability condition; its task is to ensure that every clause will be satisfied. Finally, there will be a set of triples, GC(garbage collection), which will enable to complete the matching if the previous conditions have been met.

The sets W, X and Y are defined as follows:

$$W = \{x_{ij}, \bar{x}_{ij} \mid 1 \le i \le n, 1 \le j \le m\},$$

that is, for each variable x_i and each clause C_j there is an element $x_{ij} \in W$ representing the appearance of x_i in C_j; similarly $\bar{x}_{ij}$ for $\bar{x}_i$ in C_j. [This is, clearly, a great waste, since x_i and $\bar{x}_i$ are unlikely to appear in the same clause, and only three variables can appear in each clause. However, we are not concerned here with efficient reductions; as long as it is polynomial it is acceptable, and the simpler construction is preferred.]

The sets X and Y are described piecewise,

---

*The use of 3SAT instead of SAT is unimportant. The same reduction proves also SAT $\propto$ 3DM directly.

$$X = A_1 \cup S \cup G$$

$$Y = A_2 \cup S \cup G$$

where A_i, S and G are pairwise disjoint; A_i plays a role in the assignment consistency, S in the satisfiability and G in the garbage collection.

$$A_1 = \{a_{ij} \mid 1 \le i \le n, \ 1 \le j \le m\},$$

$$A_2 = \{b_{ij} \mid 1 \le i \le n, \ 1 \le j \le m\},$$

$$S = \{s_j \mid 1 \le j \le m\},$$

$$G = \{g_k \mid 1 \le k \le m(n-1)\}.$$

Now $M = AC \cup SC \cup GC$, where AC, SC, GC are pairwise disjoint. They are defined as follows:

$$AC = \bigcup_{i=1}^{n} AC_i$$

where

$$AC_i = \{(x_{ij}, a_{ij}, b_{ij}) \mid 1 \le j \le m\} \cup$$

$$\{(\bar{x}_{ij}, a_{i,j+1}, b_{ij}) \mid 1 \le j \le m-1\} \cup \{(\bar{x}_{im}, a_{i1}, b_{im})\}.$$

The structure of AC_i is described in Figure 9.1, where the triples are circumscribed and the i index is dropped. Since each a_{ij} and b_{ij} participates in only two triples, for every i M' must contain either all triples of the type (x_{ij}, a_{ij}, b_{ij}) or all of the other type, but no mixture is possible. This represents the fact that all appearances of x_i are 'false' (corresponds to covering the a_{ij}'s and b_{ij}'s with the x_{ij}'s) or all are 'true', and if $x_i = T$, then $\bar{x}_i = F$, etc.

$$SC = \bigcup_{j=1}^{m} SC_j$$

where

$$SC_j = \{(x_{ij}, s_j, s_j) \mid x_i \in C_j\} \cup \{\bar{x}_{ij}, s_j, s_j) \mid \bar{x}_i \in C_j\}.$$

Since we are using 3SAT, each SC_j contains three triples. In order to cover s_j, as a component in the second and third dimensions, one, and only one of the triples must be in M'. Clearly this can be the case only if x_{ij} ($\bar{x}_{ij}$) has not been used in $M' \cap AC$, namely, if x_i gets 'true' ('false') assignment.

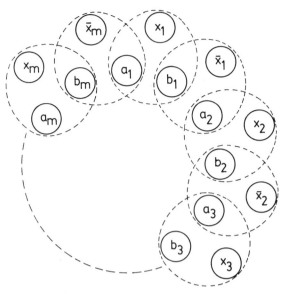

Figure 9.1

Now, $M' \cap (AC \cup SC)$ uses up exactly $mn + m$ of the $2mn$ elements of W. In order to be able to cover the remaining $m(n - 1)$ elements we have

$$GC = \{(x_{ij}, g_k, g_k), (\bar{x}_{ij}, g_k, g_k) \mid 1 \leq i \leq n, 1 \leq j \leq m, 1 \leq k \leq m(n - 1)\}.$$
Q.E.D.

The *exact cover by 3-sets* (3XC) problem is defined as follows. We are given a collection of sets $S_1, S_2, \ldots, S_n$, all subsets of some universal set U, and each consists of exactly three elements. Is there a subset I of $\{1, 2, \ldots, n\}$ such that

$$\bigcup_{i \in I} S_1 = U \tag{9.1}$$

and if $i, j \in I$ and $i \neq j$, then

$$S_i \cap S_j = \emptyset . \tag{9.2}$$

The word 'cover' describes equation (9.1), and the word 'exact' describes condition (9.2), i.e. that the sets in the cover must be disjoint. Thus, 3XC can be described, in short, as follows:

Input: A collection C of 3-sets, all subsets of a given universal set U.

Question: Is there a subcollection of C which is an exact cover of U?

Theorem 9.4: 3XC is NPC.

Proof: Obviously, 3XC ϵ *NP*. We show now that 3DM $\propto$ 3XC. Given the sets W, X, Y and $M \subseteq W \times X \times Y$ which specify the instance of 3DM, let us assume that W, X and Y are pairwise disjoint. If they are not, then by using new names we can easily change them, (and M, accordingly) to satisfy this assumption.

Define now

$$C = \{\{w, x, y\} \mid (w, x, y) \epsilon M\}$$

and

$$U = W \cup X \cup Y.$$

We claim that there is an exact covering of U by a subcollection of C if and only if there is a complete matching $M' \subseteq M$.

Clearly, if M' is a matching then

$$C' = \{\{w, x, y\} \mid (w, x, y) \epsilon M'\}$$

is an exact cover. And conversely, if C' is an exact cover, then

$$M' = \{(w, x, y) \mid \{w, x, y\} \epsilon C'\}$$

is a complete matching.

Q.E.D.

The *exact cover* (XC) problem is similar to the 3XC, except that the sets are not necessarily 3-sets. Clearly 3XC $\propto$ XC, and therefore XC is NPC too.

The *set covering* (SC) problem is defined as follows:

Input: A collection of sets C, all subsets of the universal set U, and an integer k.

Question: Is there a subcollection C' of C, such that C' is a cover of U and the number of sets in C' is less than or equal to k?

The conditions that C' must satisfy are therefore:

$$\bigcup_{S \in C'} S = U,$$

and

$$|C'| \leq k.$$

Theorem 9.5: SC is NPC.

Proof. Let us show that 3XC $\propto$ SC.

Given C and U as input to 3XC, let $k = \lfloor |U|/3 \rfloor$. Now, C, U and k constitute the input to the SC problem.

If $|U|$ is not divisible by 3 then clearly the answer to the 3XC is 'no'. Since each set covers 3 elements, k sets can cover at most $3k$ elements and therefore cannot cover U. Thus, the answer to SC is also 'no'.

If $|U|$ is divisible by 3 then if C' is an exact cover then C' is also a cover and $|C'| = |U|/3 = k$ since each set covers exactly 3 elements. Also, if C' is a cover with $|C'| \leq k$ then $|C'| = k$, since every set covers, at most, three elements and there are $|U| = 3k$ elements. It follows that if C' is such a cover then the sets in it are disjoint, and therefore the cover is exact.

$$\text{Q.E.D.}$$

We can also define the set cover problem by 3-sets (3SC), and the same proof shows that it is NPC.

Usually, when we encounter a covering problem, it is stated as an optimization problem: Given a collection C of subsets of U, find the smallest subcollection C' which covers U. Clearly, if we could solve this optimization problem then we could also solve SC in polynomial time. It follows that if SC is hard to solve, as is suggested by its being NPC, then so is the optimization problem.

PROBLEMS

9.1 A set $M \subseteq W \times X \times Y$ is *pairwise consistent* if for all a, b, c, w, x, y, $\{(a, b, y), (a, x, c), (w, b, c)\} \subset M$ implies that $(a, b, c) \in M$. Prove that 3DM remains NPC even if M is restricted to be pairwise consistent. (Hint: Review the proof of Theorem 9.3.)

9.2 The *set packing* problem is defined as follows:
   Input: A collection C of sets and a positive integer k.
   Question: Does C contain k sets which are pairwise disjoint?
   Prove that this problem is NPC.

9.3 The *hitting set* (HS) problem is defined as follows:
   Input: A collection C of subsets of a universal set U and a positive
      integer k.
   Question: Is there a $U' \subseteq U$ such that $|U'| \leq k$ and U' contains at
      least one element from each subset in C?
   Prove that HS is NPC.
   (Hint: Show that SC $\propto$ HS.
   Let $\{S_1, S_2, \ldots, S_m\}$ where $S_j \subseteq S = \{u_1, u_2, \ldots, u_n\}$ and k be the
   input to SC.
   Define $R_i = \{j \mid u_i \in S_j\}$.
   The input to HS is as follows:

$$C = \{R_i \mid 1 \leq i \leq n\},$$
$$U = \{j \mid 1 \leq j \leq m\}$$

and k remains unchanged.)

9.4 The 0-1 Knapsack problem (0-1 KNAP) is defined as follows:
   Input: Positive integers $a_1, a_2, \ldots, a_n, b$.
   Question: Is there a subset $I \subseteq \{1, 2, \ldots, n\}$ such that $\Sigma_{i \in I} a_i = b$?
   (a) Prove that 0-1 KNAP is NPC if the integers are encoded in binary
   or any other radix (greater than or equal to 2) system. (Hint: show that
   XC $\propto$ 0-1 KNAP. Let the input I of XC consist of subsets $S_1, S_2, \ldots,$
   S_m of the universal set $U = \{u_0, u_1, \ldots, u_{t-1}\}$. Define $f(I)$ as follows:

$$a_i = \Sigma_{u_j \in S_i} (m + 1)^j \qquad \text{for } i = 1, 2, \ldots, m,$$

$$n = m,$$

$$\text{and} \quad b = \sum_{i=0}^{t-1} (m + 1)^i = \frac{(m + 1)^t - 1}{m}.)$$

(b) Show that if b is expressed in the unary notation then 0-1 KNAP is
solvable in polynomial time. (Hint: prepare a binary vector $(x_0, x_1, \ldots,$
$x_b)$ in which initially $x_0 = 1$ and $x_1 = x_2 \cdots = x_b = 0$. For each
$i = 1, 2, \ldots, n$ and $j = b - 1, b - 2, \ldots, 2, 1$, if $x_j = 1, j + a_i \leq b$

and $x_{j+a_i} = 0$, set $x_{j+a_i} \leftarrow 1$. At the end, the answer to $0-1$ KNAP is 'yes' if and only if $x_b = 1$.)

9.5 The partition problem (PART) is defined as follows:

Input: Positive integers $p_1, p_2, \ldots, p_m$.

Question: Is there a subset $J \subseteq \{1, 2, \ldots, m\}$ such that

$$\sum_{i \in J} p_i = \sum_{i \notin J} p_i?$$

Prove that (in radix ≥ 2) PART is NPC.
(Hint: By 0-1 KNAP $\propto$ PART. Define

$$m = n + 2,$$

$$p_i = a_i \quad \text{for} \quad i = 1, 2, \ldots, n,$$

$$p_{n+1} = \sum_{i=1}^{n} a_i + b,$$

$$p_{n+2} = 2 \cdot \sum_{i=1}^{n} a_i - b.)$$

REFERENCES

[1] Edmonds, J., "Paths, Trees and Flowers", Canad. J. of Math., Vol. 17, 1965, pp. 449-467.

[2] Hopcroft, J. E., and Ullman, J. D., *Formal Languages and their Relation to Automata*, Addison-Wesley, 1969.

[3] Minsky, M., Computation: *Finite and Infinite Machines*, Prentice Hall, 1967.

[4] Karp, R. M. "Reducibility among Combinatorial Problems", in R. E. Miller and J. W. Thatcher (eds.), *Complexity of Computer Computations*, Plenum Press, 1972, pp. 85-104.

[5] Cook, S. A., "The Complexity of Theorem Proving Procedures", *Proc. 3rd Ann. ACM Symp. on Theory of Computing*, ACM, 1971, pp. 151-158.

[6] Aho, A. V., Hopcroft, J. E., and Ullman, J. D., *The Design and Analysis of Computer Algorithms*, Addison-Wesley, 1974.

[7] Garey, M. R., and Johnson, D. S., *Computers and Intractability; A Guide to the Theory of NP-Completeness*, W. H. Freeman, 1979.

Chapter 10

NP-COMPLETE GRAPH PROBLEMS

There are many known NPC graph problems, and we cannot possibly describe them all in one chapter. The interested reader can refer to the book of Garey and Johnson [1], for the most complete list of NPC problems. In this chapter some of the most interesting NPC graph problems are discussed.

10.1 CLIQUE, INDEPENDENT SET AND VERTEX COVER

The graphs in this section are all finite, undirected, have no parallel edges and no self-loops. These assumptions are natural when we deal with any of the problems of this section.

A *clique* of a graph $G(V, E)$ is a subset of vertices, C, such that if $u, v \in C$ then $u - v$ in G.* An *independent set* of G is a subset of vertices, S, such that if $u, v \in S$ then $u \not\sim v$ in G. (Here $u - v$ means that there is an edge connecting u and v, and $u \not\sim v$ means that there is no such edge.) A *vertex cover* (of the edges) is a subset, C, of vertices such that if $u - v$ then $\{u, v\} \cap C \neq \emptyset$.

The maximum clique problem (CLIQUE) is defined as follows:

Input: A graph $G(V, E)$ and a positive integer $k \leq |V|$.

Question: Is there a clique C in G such that $|C| \geq k$?

Clearly, this is called the maximum clique problem because if we could solve it efficiently we could also solve the problem of finding the size of a maximum clique efficiently. Also, if CLIQUE is shown to be hard to solve then the problem of finding a clique of maximum size must be hard to solve.

---

*The reader is warned that there is another common definition; according to it a clique is a maximal set with this property.

Theorem 10.1: CLIQUE is NPC.

Proof: Obviously CLIQUE $\in$ *NP*. We show now that 3DM $\propto$ CLIQUE. Let $M \subseteq W \times X \times Y$ be the input I of 3DM. Define $f(I)$ as follows:

$V = M$,

$E = \{m_1 - m_2 | m_1, m_2 \in M$ and these two triples do not have any component in common$\}$,

$k = |W|$.

The graph $G(V, E)$ has a clique of size k if and only if there is a complete matching $M' \subseteq M$.

Q.E.D.

The maximum independent set problem (IND) is defined as follows:

Input: A graph $G(V, E)$ and a positive integer $k \leq |V|$.

Question: Is there an independent set $S \subseteq V$ in G such that $|S| \geq k$.

Theorem 10.2: IND is NPC.

Proof: Let us show that CLIQUE $\propto$ IND.

If the input I of CLIQUE consists of $G(V, E)$ and k, let $f(I)$ consist of $\overline{G}$ and k, where $\overline{G}$ is the graph complementary to G; i.e., $\overline{G}(V, E')$ is defined by: $u - v$ in $\overline{G}$ if and only if $u \nleftrightarrow v$ in G. A subset of vertices $S \subseteq V$ is a clique in G if and only if S is an independent set in $\overline{G}$. Thus the answer to I, with respect to CLIQUE is 'yes' if and only if the answer to $f(I)$ is 'yes', with respect to IND.

Q.E.D.

The minimum vertex cover problem (VC) is defined as follows:

Input: A graph $G(V, E)$ and a positive integer $l \leq |V|$.

Question: Is there a vertex cover C such that $|C| \leq l$.

Theorem 10.3: VC is NPC.

Proof: Let us show that IND $\propto$ VC.

Let the input I of IND consist of a graph $G(V, E)$ and an integer k. In $f(I)$, G is unchanged and $l = |V| - k$.

If there is an independent set S such that $|S| \geq k$ then clearly $V - S$ is
a vertex cover. The number of vertices in $V - S$ is $|V| - |S| \leq |V| - k$
$= l$. Thus the answer to $f(I)$ with respect to VC is also 'yes'. Conversely, if
C is a vertex cover and $|C| \leq l$, then $V - C$ must be an independent set.
The number of vertices in $V - C$ is $|V| - |C| \geq |V| - l = k$. Thus the
answer to I with respect to IND is also 'yes'.

Q.E.D.

10.2 HAMILTON PATHS AND CIRCUITS

The *directed Hamilton path* problem (DHP) is defined as follows:

Input: A digraph $G(V, E)$ and two vertices s, t.

Question: Is there a simple directed path which starts in s, ends in t and
passes through all the other vertices?

Theorem 10.4: DHP is NPC.

Proof: Let us show that VC $\propto$ DHP. (This reduction is due to E. Lawler.)
Let $G(V, E)$ be the graph and k be the positive integer in the input I to
VC. For every vertex, let $e(v, 1)$, $e(v, 2)$, $\ldots$, $e(v, d(v))$ be the edges
incident to v in G, where $d(v)$ is the degree of v.

The digraph $G'(V', E')$ of $f(I)$ is constructed as follows:

$$V' = \{a_0, a_1, \ldots, a_k\} \cup \bigcup_{v \in V} \{(v, 1, i), (v, 2, i) | 1 \leq i \leq d(v)\}.$$

Here, $a_0, a_1, \ldots, a_k$ are new symbols, and for every vertex v of G, there
are $2 \cdot d(v)$ vertices in G'.

$$
\begin{aligned}
E' = \{a_i &\rightarrow (v, 1, 1) | 0 \leq i < k \text{ and } v \in V\} \cup \\
\{(v, 2, d(v)) &\rightarrow a_i | 0 < i \leq k \text{ and } v \in V\} \cup \\
\{(u, 1, i) &\rightarrow (v, 1, j), (u, 2, i) \rightarrow (v, 2, j) | e(u, i) = e(v, j)\} \cup \\
\{(v, 1, i) &\rightarrow (v, 2, i) | v \in V \text{ and } 1 \leq i \leq d(v)\} \cup \\
\{(v, 2, i) &\rightarrow (v, 1, i + 1) | v \in V \text{ and } 1 \leq i < d(v)\}.
\end{aligned}
$$

The last two parts, in the definition of E', describe a directed path from
$(v, 1, 1)$ to $(v, 2, d(v))$ which passes through all the other $2 \cdot d(v) - 2$ ver-

tices in G', associated with v. This path will be called v's track. For every edge $u \overset{e}{-} v$ in G there is a linkage between u's track and v's track as specified in the third part in the definition of E'; if e is the i-th edge incident to u ($e = e(u, i)$) and is the j-th edge incident to v ($e = e(v, j)$) then the connections it implies are as shown in Fig. 10.1. For every $0 \le i < k$ and every $v \in V$ by the first part of the definition of E', a_i is connected by an edge to $(v, 1, 1)$. Similarly, by the second part, for every $0 < i \le k$ and $v \in V$, $(v, 2, d(v))$ is connected by an edge to a_i. Thus, the a_i vertices serve as passages from one track to another track.

The reason for the construction of the linkage as shown in Fig. 10.1 is that if the Hamilton path enters from A it must exit from C, or else, either $(v, 1, j)$ or $(u, 2, i)$ cannot be included in the path. The path can enter from A and go through all four vertices and exit via C; it can enter from B, go through all four vertices and exit via D; it can enter both from A and B and exit via C and D, respectively. Thus, if the path enters $(u, 1, 1)$ (from some a_i), it will go through all the $2 \cdot d(u)$ vertices on u's track and exit from $(u, 2, d(u))$ (to some a_j). It may cover pairs of vertices $(v, 1, j)$, $(v, 2, j)$, in addition, if $u - v$ in G.

Let $s = a_0$ and $t = a_k$. This completes the definition of $f(I)$. We claim that G has a k-vertex cover if and only if there is a Hamilton path from a_0 to a_k in G'.

Assume $C = \{v_1, v_2, \ldots, v_k\}$ is a vertex cover of G.* One can construct a Hamilton path in G', from a_0 to a_k, as follows. Start with an edge from a_0 to $(v_1, 1, 1)$, down the v_1 track to $(v_1, 2, d(v_1))$, from there to a_1, to $(v_2, 1, 1)$, down the v_2 track to $(v_2, 2, d(v_2))$, etc. Finally, from $(v_k, 2, d(v_k))$ to a_k. Now, for every edge $u \overset{e}{-} v$ in G, if one vertex, say u, belongs to C but $v \notin C$, the vertices $(v, 1, j)$ and $(v, 2, j)$, where $e = e(v, j)$, are included by making a detour in the u track. Assume $e = e(u, i)$, then instead of going directly from $(u, 1, i)$ to $(u, 2, 1)$, insert the detour $(u, 1, i) \to (v, 1, j) \to (v, 2, j) \to (u, 2, i)$. Since C is a vertex cover, we can include in this way all the vertices on the unused tracks.

If P is an Hamilton path from a_0 to a_k in G', then we can construct a k-vertex cover, S, of G, as follows. $v \in S$ if and only if v's track is used in P. The number of vertices in S is exactly k. Consider now an edge $u \overset{e}{-} v$ in G. If both the v track and u track are used in P then clearly e is (doubly) covered. If not, the only way to have $(u, 1, i)$, $(u, 2, i)$, $(v, 1, j)$ and $(v, 2, j)$ in P is either to use the u track or the v track. Thus, e is covered by either u or v.   Q.E.D.

---

*There is no loss of generality in assuming that the cover contains exactly k vertices, for addition of vertices does not ruin a cover.

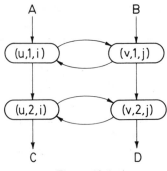

Figure 10.1

The directed Hamilton circuit problem (DHC) is defined similarly:

Input: A digraph $G(V, E)$.

Question: Is there a simple directed circuit in G, which passes through all the vertices?

Clearly, DHC is also NPC. This is easily proved by showing that DHP $\propto$ DHC. Given a digraph $G(V, E)$ and two vertices s and t, as the input I to DHP, construct the digraph $G'(V', E')$ by adding to G one new vertex a and two edges, $t \to a$ and $a \to s$. Clearly, there is a Hamilton circuit in G' if and only if there is a Hamilton path from s to t in G.

The (undirected) Hamilton path problem (HP) is also natural:

Input: A graph $G(V, E)$ and two vertices s, t.

Question: Is there a simple path in G with endpoints s and t, which passes through all the other vertices?

Theorem 10.5: HP is NPC.

Proof: Let us show that DHP $\propto$ HP. (This proof is due to R. E. Tarjan.) Let the input I of DHP consist of the digraph $G(V, E)$ and the two vertices be s and t. The graph $G'(V', E')$, of $f(I)$, is defined as follows:

$V' = \{(v, 0), (v, 1), (v, 2) \mid v \in V\}$

$E' = \{(v, 0) - (v, 1), (v, 1) - (v, 2) \mid v \in V\} \cup \{(u, 2) - (v, 0) \mid u \to v \text{ in } G\}$.

The two endpoints, of $f(I)$, are $(s, 0)$ and $(t, 2)$.

It is easy to see that if there is a directed Hamilton path P, from s to t, in G, then there is a Hamilton path P', in G', with endpoints $(s, 0)$ and $(t, 2)$: P' consists of all the edges in the first part of the definition of E', and if $u \rightarrow v$ in P then $(u, 2) - (v, 0)$ is included in P'.

Assume now that P' is a solution to the HP problem specified by $f(I)$. Clearly, all the edges of the first part of the definition of E' appear in P', or there would not be any way to pass through some $(v, 1)$. We can now scan P' from $(s, 0)$ to $(t, 2)$. Clearly, P' starts with $(s, 0) - (s, 1) - (s, 2)$ and ends with $(t, 0) - (t, 1) - (t, 2)$. Whenever $(u, 2) - (v, 0)$ is used in P', we can use $u \rightarrow v$ in G. Thus, the resulting directed path is simple, starts in s, ends in t, and passes through all other vertices of G.

Q.E.D.

The (undirected) Hamilton circuit problem (HC) is also NPC, and this is again proved by showing that HP $\propto$ HC. Again the reduction is simply by adding a new vertex a and two new edges $t - a$ and $a - s$.

The *traveling salesman* problem, is really not one problem. Generally, a graph or a digraph is given, with length assigned to the edges. The problem is to find a minimum circuit, or path from a vertex s to a vertex t, such that every vertex is on it. Vertices may, or may not, be more than once on the circuit or path.

For definiteness, let us assume that $G(V, E)$ is an undirected graph, each $e \in E$ is assigned a length $l(e)$ and we are required to find a simple circuit which passes through all the vertices and whose sum of edge lengths is minimum. Clearly, if we could solve this traveling salesman problem in polynomial time we could also solve HC. Simply, assign length 1 to all the edges and solve the traveling salesman problem. This observation remains valid even if in the traveling salesman problem the circuit is not required to be simple; a minimum circuit is of length $|V|$ if and only if it is Hamiltonian. Similarly, the directed versions are related. We conclude that the traveling salesman problems are hard to solve if $P \neq NP$.

10.3 COLORING OF GRAPHS

One of the classical problems in graph theory is that of coloring the vertices of a graph in such a way that no two adjacent vertices (i.e. connected by an edge) are assigned the same color. The minimum number of colors necessary to color G is called the *chromatic number*, $\gamma(G)$, of G.

In this section, we shall show that this problem is NPC. The problem remains NPC even if all we ask is whether $\gamma(G) \le 3$. Furthermore, even if we restrict the question to planar graph, the problem remains NPC. Even if we restrict the problem to a class of planar graphs with well behaved planar realization, the problem of whether $\gamma(G) \le 3$ is still NPC. One such definition for well behaved realization is that all edges are straight lines, no angle is less than $10°$ and the edge lengths are in between two given bounds.

First we consider the 3-Coloration problem, (3C), which is defined as follows:

Input: A graph $G(V, E)$.

Question: Can one assign each vertex a color, so that only three colors are used and no two adjacent vertices are assigned the same color? (In short: Is $\gamma(G) \le 3$?)

Theorem 10.6: 3C is NPC.

Proof: We show that 3SAT $\propto$ 3C. (The proof of this theorem, and the next, follows the works of Stockmeyer [2] and Garey, Johnson and Stockmeyer [3].) Let the set of literals, of the input I to 3SAT, be $\{x_1, x_2, \ldots, x_n, \bar{x}_1, \bar{x}_2, \ldots, \bar{x}_n\}$ and the clauses be $C_1, C_2, \ldots, C_m$.

The graph $G(V, E)$, which is the input $f(I)$ to 3C is defined as follows:

$$V = \{a, b\} \cup \{x_i, \bar{x}_i \,|\, 1 \le i \le n\} \cup \{w_{ij} \,|\, 1 \le i \le 6 \text{ and } 1 \le j \le m\}.$$

$$E = \{a - b\} \cup \{a - x_i, a - \bar{x}_i, x_i - \bar{x}_i \,|\, 1 \le i \le n\} \cup$$
$$\{w_{1j} - w_{2j}, w_{1j} - w_{4j}, w_{2j} - w_{4j}, w_{4j} - w_{5j},$$
$$w_{3j} - w_{5j}, w_{3j} - w_{6j}, w_{5j} - w_{6j}, w_{6j} - b \,|\, 1 \le j \le m\} \cup$$
$$\{\xi_{1j} - w_{1j}, \xi_{2j} - w_{2j}, \xi_{3j} - w_{3j} \,|\, 1 \le j \le m \text{ and }$$
$$C_j = \{\xi_{1j}, \xi_{2j}, \xi_{3j}\}\}.$$

The structure of the last two parts in the definition of E, for each j, is shown in Figure 10.2.

The significance of this structure is as follows: Assume each of the vertices ξ_{1j}, ξ_{2j} and ξ_{3j} is colored 0 or 1, (we assume the three colors are 0, 1 and 2) and ignore for the moment vertex b. We claim that w_{6j} can be colored 1 or 2 if and only if not all three vertices ξ_{1j}, ξ_{2j} and ξ_{3j} are colored 0. First, it is easy to see that if ξ_{1j}, ξ_{2j}, ξ_{3j} are colored 0 then w_{4j} must also

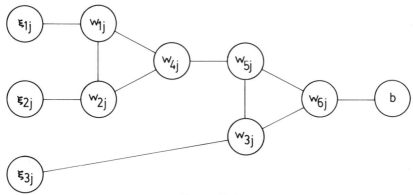

Figure 10.2

be colored 0, and therefore w_{6j} must be colored 0. But, as the reader may check for himself, if at least one of ξ_{1j}, ξ_{2j}, ξ_{3j} is colored 1 then w_{6j} can be colored 1.

The structure of the first two parts of E's definition is shown in Fig. 10.3. Clearly, if a is colored 2 then all the literal-vertices must be colored 0 and 1, one of these colors is used for x_i and the other for $\bar{x}_i$. Assume I is satisfiable by some truth value assignment to the literals. To see that $f(I)$ is 3-colorable, assign a the color 2. Assign the literal ξ the color 1 if it is 'true' and 0 if it is 'false'. Now, since no triple ξ_{1j}, ξ_{2j}, ξ_{3j} is assigned all zeros, we can color w_{1j}, w_{2j}, ..., w_{6j} in such a way that w_{6j} is colored 1, for all $j = 1, 2, \ldots, m$. Thus, b is colorable 0, and the 3-coloration of G is complete. Conversely, if G is 3-colorable, call a's color 2 and b's color 0. Clearly, all the literal vertices are colored 0 and 1, and w_{6j} cannot be colored 0. Thus, for each triple ξ_{1j}, ξ_{2j}, ξ_{3j}, not all three are colored 0. Now, if we assign a literal 'true' if and only if its corresponding vertex is colored 1, the assignment satisfies all clauses.

Q.E.D.

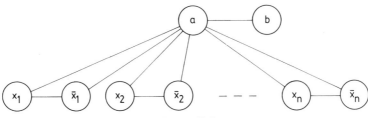

Figure 10.3

Before we turn to the problem of 3-colorability of planar graphs in general let us consider the diamond shape planar graph D of Figure 10.4. Assume we color it with the three colors 0, 1, and 2. If u_0 is colored 0, the circuit $u_1 - u_2 - u_3 - u_4 - u_1$ is colored 1 and 2, alternately. For definiteness, assume u_1 and u_3 are colored 1 and u_2 and u_4 are colored 2. Now, there are two possible colors for u_5; i.e. 0 or 2. If u_5 is colored 0, then v_1 is colored 2, v_4 is colored 1, u_6 is colored 0, v_2 is colored 2, u_7 is colored 0, v_3 is colored 1 and u_8 is colored 0. The important fact is that v_1 and v_2 have the same color (2) and v_3 and v_4 have the same color (1). If u_5 is colored 2, then v_1 is colored 0, u_8 is colored 1, v_3 is colored 0, u_7 is colored 2, v_2 is colored 0, u_6 is colored 1 and v_4 is colored 0; i.e. all four vertices, v_1, v_2, v_3 and v_4 are colored identically. We conclude that in every 3-coloring of D v_1 and v_2 must have the same color, v_3 and v_4 must have the same color, but the color of v_1 and v_3 may or may not be the same. Thus, D, effectively

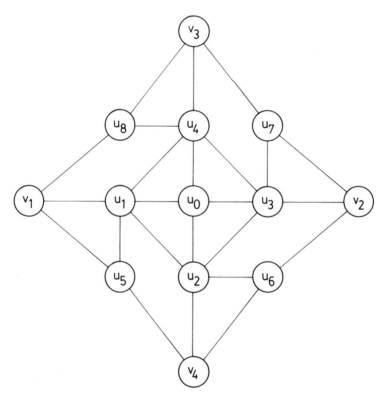

Figure 10.4

performes a crossover of the coloring of v_1 to v_2, of v_3 to v_4, while no constraint as to the equality or inequality of the colors of v_1 and v_3 is introduced.

The 3-Coloration of planar graphs problem (3CP) is defined as follows:

Input: A plane graph $G(V, E)$.

Question: Is $\gamma(G) \leq 3$?

(We remind the reader that a plane graph is a drawing of a graph in the plane, so that no two edges share a point except, possibly, a mutual endpoint.)

Theorem 10.7: 3CP is NPC.

Proof: We show that 3C $\propto$ 3CP. Let $G(V, E)$ be the input I to the 3C problem, where $V = \{v_1, v_2, \ldots, v_n\}$. We construct $f(I)$ in two steps. First, construct a general layout, as demonstrated in Fig. 10.5 for the case of $n = 5$. This general layout depends only on n, and not on E. (The idea here is a variation of Johnson's construction, based on a simple and well known sorting network, from which the last two layers are omitted. For a descrip-

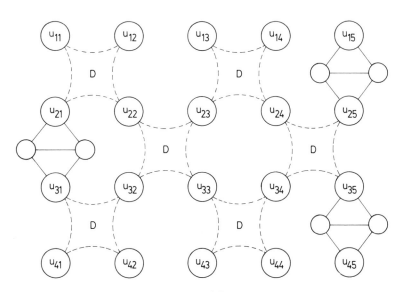

Figure 10.5

tion of the sorting network see, for example, references [4] or [5].) This layout has $n - 1$ main layers of vertices and n main columns of vertices; these vertices are denoted u_{ij}, where $1 \leq i \leq n - 1$ and $1 \leq j \leq n$. If $i + j$ is even then a copy of D is constructed with vertices u_{ij}, $u_{i(j+1)}$, $u_{(i+1)j}$ and $u_{(i+1)(j+1)}$ playing the role of D's four points. If $i < n - 1$ and even then u_{i1} and $u_{(i+1)1}$ are connected via two new vertices as shown in Fig. 10.5 in the case of u_{21} and u_{31}. If $i < n - 1$ and $i + n$ is even then u_{in} is connected, similarly, to $u_{(i+1)n}$; see u_{15} and u_{25}, and also u_{35} and u_{45} in Fig. 10.5. This completes the description of the layout.

Assuming all vertices* are colored using only three colors. Clearly, u_{11}, u_{22}, ..., $u_{(n-1)(n-1)}$ are all colored identically. If $i > 1$ and odd then u_{1i}, $u_{2(i+1)}$, ..., $u_{(n-i+1)n}$, $u_{(n-i+2)n}$, $u_{(n-i+3)(n-1)}$, ..., $u_{(n-1)(n-i+3)}$ are all colored identically. Also, if n is even, u_{1n}, $u_{2(n-1)}$, ..., $u_{(n-1)2}$ is such a track, and if $i < n$ and even then u_{1i}, $u_{2(i-1)}$, ..., u_{i1}, $u_{(i+1)1}$, $u_{(i+2)2}$, ..., $u_{(n-1)(n-i-1)}$ is such a track. Let us call the track that begins at u_{1i}, the i-th track.

We want to show that for every two tracks there is at least one $1 \leq i \leq n - 1$ and $1 \leq j < n$ such that u_{ij} is on one of these tracks and $u_{i(j+1)}$ is on the other. If this happens we say that these two tracks are adjacent at the i-th level.

Assume i and j are both odd and $i < j$. If $i = 1$ then its track cuts diagonally from u_{11} to $u_{(n-1)(n-1)}$, and clearly it is adjacent to every other track somewhere; in fact, if $j \neq 3$ then the two tracks intersect. If $i > 1$ then the i-th track ends in $u_{(n-1)(n-i+3)}$ and the j-th in $u_{(n-1)(n-j+3)}$. Thus, they intersect somewhere.

The case of both i and j even is handled similarly.

If $i < j$, i is odd and j is even, then the two tracks intersect before either one reaches the side.

If $i < j$, i is even, j is odd, then the i-th track ends in $u_{(n-1)(n-i-1)}$ and the j-th track ends in $u_{(n-1)(n-j+3)}$. If $j - i = 1$ they are adjacent at the first level. If $j - i \geq 3$ then

$$(n - j + 3) - (n - i - 1) = -(j - i) + 4 \leq 1,$$

and if the two tracks do not intersect they are adjacent at the $(n - 1)$-st level.

Now, we turn to the second part of the reduction. If $v_i - v_j$ in G, add an edge $u_{kl} - u_{k(l+1)}$ to the layout, where u_{kl} is on the i-th track (or the j-th

---

*The number of vertices in the layout is altogether $n(n - 1) + (9/2) \cdot (n - 1) \cdot (n - 2) + 2(n - 2)$.

track) and $u_{k(l+1)}$ is on the j-th track (or the i-th track). Such a k and l can be found since every two tracks are adjacent somewhere.

Now, if G is 3-colorable, color all the vertices of the i-th track with the color of v_i. Clearly, the remaining vertices of $f(I)$ can also be legally colored; the vertices in the D structures and the connecting pairs are easily colored, and the edges which correspond to edges of E connect between tracks of different color. Conversely, if $f(I)$ is 3-colorable, each track is uniformly colored, and we can assign v_i, in G, the same color. No two adjacent vertices in G get the same color, because there is an edge connecting the two corresponding tracks in $f(I)$, and since $f(I)$ is legally colored, the colors of these two tracks are different.

Q.E.D.

As the reader can easily observe the plane graph, $f(I)$, constructed in this proof is "well behaved". This justifies the claim, as made earlier in the section, that even when the 3-colorability problem is restricted to such graphs, the problem remains NPC.

10.4 FEEDBACK SETS IN DIGRAPHS

In certain engineering applications of graph theory one is given a digraph which contains directed circuits, and one wants to eliminate from the digraph a minimum number of elements so that the remaining digraph by acyclic; i.e. contain no directed circuits. We show (following Karp [7]) that both cases, that of eliminating vertices, and of eliminating edges are NPC.

The minimum feedback vertex set problem (FVS) is stated as a decision problem as follows:

Input: Digraph $G(V, E)$ and a positive integer k.

Question: Is there a subset of vertices V' such that $|V'| \leq k$ and the digraph resulting from G by eliminating all the vertices in V' and their incident edges is acyclic?

Theorem 10.8: FVS is NPC.

Proof: We show that VC $\propto$ FVS. Let the input I to the vertex cover problem consist of the graph $G(V, E)$ and the integer k. The input $f(I)$ to FVS consists of a digraph $H(V, F)$ and the same integer k, where F is defined as follows:

$$F = \{a \rightarrow b, \ b \rightarrow a \,|\, a - b \ \text{in} \ E\}.$$

Since each edge $a - b$ of G corresponds to a directed circuit $a \rightarrow b \rightarrow a$ of H, clearly, a feedback vertex set S must contain either a or b (or both). Thus the set S is a vertex cover of all the edges of G. Also, if S is a vertex cover of G, the elimination of S and all edges incident to its elements from H, leaves no edges, and therefore no directed circuits. Thus, G has a vertex cover with k or less vertices, if and only if H has a feedback vertex set of k or less vertices.

Q.E.D.

The minimum feedback edge set problem (FES) is defined as follows:

Input: A digraph $G(V, E)$ and a positive integer k.

Question: Is there a subset of edges E', such that $|E'| \leq k$ and $G'(V, E - E')$ is acyclic?

Theorem 10.9: FES is NPC.

Proof: We show that FVS $\propto$ FES. Let the input I to FVS consist of a digraph $G(V, E)$ and a positive integer k. The input $f(I)$ to FES consists of a digraph $H(W, F)$ and the same integer k, where H is defined as follows:

$$W = \{(v, 1), (v, 2) \,|\, v \in V\}$$

$$F = \{(v, 1) \rightarrow (v, 2) \,|\, v \in V\} \cup \{(u, 2) \rightarrow (v, 1) \,|\, u \rightarrow v \ \text{in} \ G\}.$$

(This reduction is similar to the technique of Problem 5.3.)

Let us call the edges of H, of the type $(v, 1) \rightarrow (v, 2)$, internal, and those of type $(u, 2) \rightarrow (v, 1)$, external. All the external edges incident to $(v, 1)$ are incoming, and there is exactly one internal edge incident to $(v, 1)$ and it is outgoing. It follows that if in a feedback edge set there is an external edge, $(u, 2) \rightarrow (v, 1)$, then it can be replaced by the internal edge, $(v, 1) \rightarrow (v, 2)$, since all the directed circuits which go through $(u, 2) \rightarrow (v, 1)$, go also through $(v, 1) \rightarrow (v, 2)$. Thus, if there is a feedback edge set F' in H, which satisfies $|F'| \leq k$, we can assume that it consists entirely of internal edges, and the set of vertices, V', in G, which correspond to these internal edges, is a feedback vertex set of G. Also, if V' is a feedback vertex set of G then the set of internal edges in H which correspond to vertices of V', is a feedback edge set of H.

Q.E.D.

10.5 STEINER TREE

The Steiner tree problem in graphs (ST) is a generalization of the minimum spanning tree problem (see Section 2.2).

Input: A connected graph $G(V, E)$, a subset of vertices $X(\subseteq V)$, a length function $l(e) > 0$ defined on the edges and a positive integer k.

Question: Is there a tree $T(W, F)$ in G, such that $X \subseteq W \subseteq V, F \subseteq E$ and $\Sigma_{e \in F} l(e) \le k$?

Clearly, if $X = V$, then ST is just the minimum spanning tree problem, stated as a decision problem. As we saw in Section 2.2, this case is solvable in polynomial time.

Theorem 10.10: ST is NPC.

Proof: (Following Karp [7].) We show that $3XC \propto ST$. Let the input I to $3XC$ consist of a universal set $U = \{u_1, u_2, \ldots, u_t\}$ and a collection $C = \{S_1, S_2, \ldots, S_n\}$ such that $S_i \subseteq U$ and $|S_i| = 3$. The input $f(I)$ to ST is defined as follows:

$$V = \{v_0\} \cup C \cup U,$$

$$E = \{v_0 \overset{e_i}{\longrightarrow} S_i | 1 \le i \le n\} \cup \{S_i \overset{f_{ij}}{\longrightarrow} u_j | u_j \in S_i\},$$

$$X = \{v_0\} \cup U,$$

$$l(e) = 1 \text{ for all } e \in E,$$

$$k = \frac{4}{3} t.$$

Assume J is a subset of $\{1, 2, \ldots, n\}$ which defines an exact cover of U; i.e., $\cup_{i \in J} S_i = U$ and if $i, j \in J$ then $S_i \cap S_j = \emptyset$. Clearly, $|J| = t/3$. Define a Steiner tree $T(W, F)$ as follows:

$$W = \{v_0\} \cup \{S_i | i \in J\} \cup U,$$

$$F = \{e_i | i \in J\} \cup \{f_{ij} | i \in J \text{ and } u_j \in S_i\}.$$

Clearly, this is a Steiner tree and its cost is exactly $4t/3$.

Now, assume $T(W, F)$ is a Steiner tree of $f(I)$. First, observe that we can assume that in T, each vertex u_j is a leaf; for if $d(u_j) > 1$ we can reduce its degree, without increasing the degree of any other u_j, and without changing T's total length, as follows. If in T u_j is connected by edges to S_{i_1} and S_{i_2}, delete the edge $f_{i,j}$ (connecting u_j with S_{i_1}). One of S_{i_1} and S_{i_2} is now disconnected from v_0; add the edge which connects it directly to v_0, to restore the connectivity.

We can now prove that if $J = \{i \mid S_i \in W\}$ then it defines an exact cover of U. Clearly, each $S_i \in W$ can have at most 3 edges in T which lead to elements of U, and since $U \subseteq W$, if any $S_i \in W$ has less then 3 such edges, $|J| > t/3$. Also, for every $S_i \in W$, $e_i \in F$. Thus, if $|J| > t/3$ then $|F| > 4t/3$ contradicting the requirement that $|F| \leq 4t/3$. Therefore, each $S_i \in W$ has exactly 3 edges to elements of U, no u_j is connected by an edge to more than one S_i and we conclude that J defines an exact cover of U.

Q.E.D.

Note that we have proved that even if all edge lengths are 1, ST remains NPC. For NP-Completeness results of the Steiner problem on a rectilinear grid and other related problems, see Garey, Graham and Johnson [8].

10.6 MAXIMUM CUT

As we saw in Chapter 5, the problem of finding a minimum cut is solvable in polynomial time. Karp [7] showed that the maximum cut problem (MAXC) is NPC. In this section we shall show that even if all the edge weights are 1 the problem is NPC. We follow Even and Shiloach [9]. Another proof of the same result was given by Garey, Johnson and Stockmeyer [3].

MAXC is defined as follows:

Input: A graph $G(V, E)$ and a positive integer k.

Question: Is there a subset S of vertices such that $|(S; \overline{S})| \geq k$.

As in Chapter 5, $(S; \overline{S})$ is the set of edges in G such that one of their endpoints is in S and the other in $\overline{S}(= V - S)$.

Theorem 10.11: MAXC is NPC.

Proof: Let us show that 3SAT $\propto$ MAXC. The reduction is done in two steps. In the first step we shall assign weights to the edges, and will reduce

3SAT to a maximum weighted cut problem. In the second step we shall show how these weights, being polynomially bounded by $|V|$, can be eliminated.

Let $C_1, C_2, \ldots, C_m$ be the clauses of 3SAT, each consisting of exactly 3 literals; the set of literals being $L = \{x_1, x_2, \ldots, x_n, \bar{x}_1, \bar{x}_2, \ldots, \bar{x}_n\}$. We construct the input to the weighted maximum cut problem, consisting of $G'(V', E')$, a weight function $w(e)$ and a positive integer k' as follows:

$$V' = \{v_i \mid 0 \le i \le m\} \cup L.$$

For every $1 \le i \le m$ let

$$A_i = \{v_0\} \cup \{v_i\} \cup C_i.$$

Now,

$$E' = \{u - v \mid u \ne v \text{ and } \exists i \ni u, v \in A_i\} \cup \{x_j - \bar{x}_j \mid 1 \le j \le n\}.$$

The weight function w is defined by

$$w(v_0 - \xi) = \sum_{i=1}^{m} |C_i \cap \{\xi\}| \quad \text{if} \quad \xi \in L,$$

$$w(\xi' - \xi'') = \sum_{i=1}^{m} |C_i \cap \{\xi'\}| \cdot |C_i \cap \{\xi''\}| \quad \text{if} \quad \xi', \xi'' \in L \text{ and } \xi' \ne \xi'',$$

$$w(x_j - \bar{x}_j) = 10 \cdot m + 1,$$

$$w(v_i - u) = 1 \quad \text{if} \quad i > 0 \text{ and } u \in A_i,$$

$$k' = (10 \cdot m + 1)n + 6 \cdot m.$$

In this construction, $|V'| = 2n + m + 1$. The number of edges, $|E'|$, is bounded by $3n + 7m$. Every clause C_i is represented by a clique A_i. In addition, there is an edge between x_j and $\bar{x}_j$ whose weights is $10 \cdot m + 1$. (This weight is designed to overweight the sum of all edges of the other type.) The weight of an edge $v_0 - \xi$ is equal to the number of times the literal ξ appears in the clauses. The weight of an edge $\xi' - \xi''$ is equal to the number of clauses in which both literals appear. The weight of all edges incident to v_i, $i > 0$, is 1. The sum of weights of all these three classes of edges is exactly $10 \cdot m$.

We now claim that the answer to the instance of 3SAT problem is the same as to the question: Is there a set $S' \subseteq V'$ such that

$$\sum_{e \in (S';\overline{S'})} w(e) \geq k' ?$$

First, assume the answer to the 3SAT problem is affirmative, and let τ be the set of literals which have a 'true' value in a consistent assignment which satisfies all the clauses. Let $\tau \subset S'$ and $L - \tau \subset \overline{S'}$. Clearly, for each $1 \leq j \leq n$, $x_j - \overline{x}_j$ belongs to the cut, and contributes $10 \cdot m + 1$ to its weight. Altogether we already have $(10 \cdot m + 1) \cdot n$ contributed to the total weight of the cut. Now, put $v_0 \in \overline{S'}$. It is convenient to interpret the rest of the edges and their weights as follows: Each A_i is a clique, and each edge appears as many times as it belongs to such cliques, which is exactly equal to its defined weight. In each of these m cliques there is at least one literal-vertex which is in S', v_0 is in $\overline{S'}$ and v_i can be put in S' or $\overline{S'}$ in such a way that 2 of A_i's vertices will be on one side and the remaining 3 on the other side. Thus, the clique contributes 6 to the cut, and the m cliques contribute together $6 \cdot m$ to the weight of the cut.

The argument above shows also that the total weight of a cut cannot exceed k', and if the answer to the weighted cut problem is affirmative then all edges of the type $x_j - \overline{x}_j$ must in the cut and, each of the m cliques must contribute 6, which is the maximum any such clique can contribute. Therefore, in each clique 2 of the vertices are on one side of the cut, and the remaining 3 are on the other side. Now, let us call the side v_0 is in, the 'false' side, and the other side, the 'true' side. It follows that at most 2 literal-vertices can be on the 'false' side. Thus, the defined assignment is consistent and satisfies all clauses.

Now, in the second step we want to reduce the weighted cut problem into an unweighted one. Actually, this has already been done above, when each edge of weight w is replaced by w parallel edges of weight 1. Since all weights are polynomially bounded by the input length to the 3SAT problem, this increase in the number of edges is polynomially bounded. However, we wish to show that the reduction can be done even if the graph is required to be simple; i.e. have no parallel edges.

Let us replace each edge $u \xrightarrow{e} v$ of G', of weight $w(e)$, by the construct shown in Fig. 10.6, where $a_1, a_2, \ldots, a_{w(e)}$ are new vertices. We claim that the new graph, G, has a cut of size $k = 2[(10 \cdot m + 1) \cdot n + 10 \cdot m] + k'$ if and only if G' has a weighted cut of size k'. This can be shown as fol-

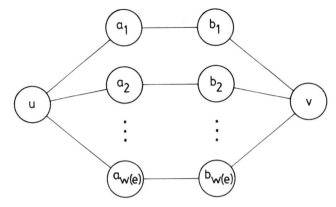

Figure 10.6

lows. In each path $u - a_l - b_l - v$ if u and v are on the same side of the cut, at most 2 edges can be in the cut. But if u and v are on different sides of the cut, by putting b_l on u's side, and a_l on v's side, the path contributes 3. Thus, each edge e of G' can give rise to at least $2 \cdot w(e)$ edges in a cut of G even if its two endpoints are on the same side. Altogether, we get this way, for all the edges of G' a contribution of $2 \cdot \Sigma_{e \in E'} w(e)$ to a cut in G, and

$$\sum_{e \in E'} w(e) = (10 \cdot m + 1)n + 10 \cdot m.$$

If the weight of $(S'; \overline{S'})$ in G' is equal to k', let $S' \subset S$ and $\overline{S'} \subset \overline{S}$. If u and v are the same side of the cut, we can assign a_l and b_l so that $u - a_l - b_l - v$ contributes 2, if not, it can contribute 3. Thus, we can complete the definition of S and $\overline{S}$ in such a way that the value of the cut $(S; \overline{S})$ in G is

$$2[(10 \cdot m + 1)n + 10 \cdot m] + k',$$

and therefore is equal to k.

Conversely, assume now that a cut $(S; \overline{S})$ of G has at least k edges. For every edge $u \overset{e}{-} v$ of G', if u and v are on the same side of this cut, e's construct contributes $2 \cdot w(e)$ at most to the cut, and without loss of generality (or change S) we can assume that this is exactly the construct's contribution. For every edge $u \overset{e}{-} v$ of G', if u and v are on different sides of the cut,

e's construct contributes as most $3 \cdot w(e)$ to the cut, and without loss of generality (or again, change S), we can assume that this is exactly the construct's contribution. Thus,

$$|(S; \overline{S})| = 2 \sum_{e \in E'} w(e) + \sum_{e \in (S'; \overline{S'})} w(e)$$

where $S' = S \cap V'$. Since $|(S; \overline{S})| = k$, we get that

$$\sum_{e \in (S'; \overline{S'})} w(e) = k'.$$

Thus, G' has a cut of weight k' if and only if G has a cut with k edges.

Q.E.D.

10.7 Linear Arrangement

In many applications of graph theory, such as printed circuits or industrial plant design, the question of optimal arrangement of its components arises. Our purpose is to show that a very simple arrangement problem is already NPC.

Let $G(V, E)$ be a simple graph (i.e. no self-loops and no parallel edges). We wish to place the vertices in the locations $1, 2, \ldots, |V|$ of a one dimensional space, one vertex in each of the locations. If u is placed in $p(u)$ and v is placed in $p(v)$, then the length to which $u - v$ stretches is $|p(u) - p(v)|$. The total length of G, in placement p, is therefore

$$\sum_{u-v \text{ in } G} |p(u) - p(v)|.$$

Our goal is to show that the problem of finding an arrangement for which the length of G is minimum, or rather the corresponding decision problem, MINLA, is NPC. As an intermediate result, we shall prove that MAXLA, defined below, is NPC.

Definition of MAXLA:

Input: A simple graph $G(V, E)$ and a positive integer k.

Question: Is there a 1-1 onto function $p: V \to \{1, 2, \ldots, |V|\}$ such that

$$\sum_{u-v \text{ in } G} |p(u) - p(v)| \geq k?$$

Lemma 10.1: MAXLA is NPC.

Proof: Let us show that MAXC $\propto$ MAXLA. (Here again, we follow [9]; a similar proof appears in [3].)

Let the input I to MAXC consist of a graph $G'(V', E')$ and a positive integer k'. We define the input $f(I)$ to MAXLA, as follows:

$$V = V' \cup \{x_1, x_2, \ldots, x_{n^3}\},$$

$$E = E'.$$

$$k = k' \cdot n^3,$$

where $n = |V'|$, and $x_1, x_2, \ldots, x_{n^3}$ are new isolated vertices. If we interpose an isolated vertex between u and v, assuming $u - v$, then $|p(u) - p(v)|$ is increased by 1.

Assume the answer to MAXC, with respect to I, is affirmative; i.e. $|(S'; \overline{S'})| \geq k'$. Define a placement p for G, which satisfies the following constraints:

If $v \in S'$ then $1 \leq p(v) \leq |S'|$.

If $v \in \{x_1, x_2, \ldots, x_{n^3}\}$ then $|S'| < p(v) \leq |S'| + n^3$.

If $v \in \overline{S'}$, then $|S'| + n^3 < p(v) \leq n + n^3$.

The number of edges which span over the n^3 vertices interposed in between S' and $\overline{S'}$ is $|(S'; \overline{S'})|$, and each has a length which exceeds n^3. Thus,

$$\sum_{u-v \text{ in } G} |p(u) - p(v)| > |(S'; \overline{S'})| \cdot n^3 \geq k' \cdot n^3 = k,$$

and the answer to $f(I)$, with respect to MAXLA, is affirmative too.

Now assume that the answer to $f(I)$, with respect to MAXLA, is affirmative. Thus, a placement $p: V \to \{1, 2, \ldots, n + n^3\}$ exists for which

$$\sum_{u-v \text{ in } G} |p(u) - p(v)| \geq k.$$

For each $1 \leq i < |V|$ define the set

$$S_i = \{v \mid v \in V \text{ and } p(v) \leq i\}.$$

Clearly $(S_i; \overline{S}_i)$ is a cut. Define j by

$$|(S_j; \overline{S}_j)| = \operatorname*{Max}_i |(S_i; \overline{S}_i)|.$$

We rearrange the vertices of G as follows. All the vertices in V' remain in their relative positions but the vertices $x_1, x_2, \ldots, x_{n^3}$ are all moved to interpose between $S_j \cap V'$ and $\overline{S}_j \cap V'$. Clearly, each x_q contributes now the maximum value to the total length of G. Thus, the new placement, p', satisfies:

$$\sum_{u-v \text{ in } G} |p'(u) - p'(v)| \geq \sum_{u-v \text{ in } G} |p(u) - p(v)| \geq k' \cdot n^3.$$

The total length of G, in placement p', can be divided into two parts:

(1) $n^3 \cdot |(S; \overline{S})|$, where $S = S_j \cap V'$ and $(S; \overline{S})$ is the cut in G' defined by S. This is the length caused by the interposition of $x_1, x_2, \ldots, x_{n^3}$ in between S and $\overline{S}$.
(2) The total length of the edges if $x_1, x_2, \ldots, x_{n^3}$ were dropped and the gap between S and $\overline{S}$ was closed. This length is clearly bounded by

$$(n - 1) \cdot 1 + (n - 2) \cdot 2 + \cdots + 1 \cdot (n - 1) = \frac{n(n^2 - 1)}{6}.$$

Thus,

$$n^3 \cdot |(S; \overline{S})| + \frac{n(n^2 - 1)}{6} \geq k' \cdot n^3,$$

or

$$|(S; \overline{S})| + \frac{n^2 - 1}{6n^2} \geq k'.$$

Since $|(S; \overline{S})|$ and k' are integers, this implies that

$$|(S; \overline{S})| \geq k'.$$

Therefore, the answer to I, with respect to MAXC is affirmative too.

Q.E.D.

Let us repeat the definition of MINLA:

Input: A simple graph $G(V, E)$ and a positive integer k.

Question: Is there a 1-1 onto function $p: V \rightarrow \{1, 2, \ldots, |V|\}$ such that

$$\sum_{u-v \text{ in } G} |p(u) - p(v)| \leq k?$$

Theorem 10.12: MINLA is NPC.

Proof: We show that MAXLA $\propto$ MINLA. This reduction is very simple. Let $G(V, E)$ and k be the input I to MAXLA. The input $f(I)$ to MINLA is defined as follows:

$$V' = V$$
$$E' = \{u - v \,|\, u \neq v \quad \text{and} \quad u \not\vdash v \text{ in } G\},$$
$$k' = n(n^2 - 1)/6 - k,$$

where $n = |V|$. Clearly, $G'(V', E')$ is the graph complementary to G. For every placement p

$$\sum_{u-v \text{ in } G} |p(u) - p(v)| + \sum_{u-v \text{ in } G'} |p(u) - p(v)| = \frac{n(n^2 - 1)}{6}.$$

Thus, $\sum_{u-v \text{ in } G} |p(u) - p(v)| \geq k$ if and only if

$$\sum_{u-v \text{ in } G'} |p(u) - p(v)| \leq k'.$$

The answer to I with respect to MAXLA is therefore identical to the answer to $f(I)$ with respect to MINLA.

Q.E.D.

10.8 MULTICOMMODITY INTEGRAL FLOW

It is our purpose to show that the multicommodity flow problem is NPC even if the number of commodities is just 2. This is true, for integral flow, both in the directed and the undirected cases. We follow here Even, Itai and Shamir [10]. It is interesting to note that in the undirected case, if flow in halves (an integral multiple of ½) is allowed, the problem is solvable in polynomial time [11].

Let $G(V, E)$ be a digraph, with a capacity function $c(e)$ (>0) defined on the edges. Vertices s_1, s_2, t_1, t_2 (not necessarily distinct) play a special role: s_1 and s_2 are called *sources* and t_1 and t_2 are called *sinks*. This data specifies the *network*.

A two-commodity flow in the network is defined by two functions $f_1(e)$ and $f_2(e)$, defined on the edges, which satisfy the following conditions:

(1) For every $e \in E, f_1(e) \geq 0, f_2(e) \geq 0$ and

$$f_1(e) + f_2(e) \leq c(e).$$

(2) For each commodity $i \in \{1, 2\}$ and every vertex $v \in V\text{-}\{s_i, t_i\}$

$$\sum_{e \in \alpha(v)} f_i(e) = \sum_{e \in \beta(v)} f_i(e).$$

(We remind the reader that $\alpha(v)$ ($\beta(v)$) is the set of edges which enter v (emanate from v) in G.)

The total flows, F_1 and F_2, of the flow functions f_1 and f_2, are defined by

$$F_i = \sum_{e \in \alpha(t_i)} f_i(e) - \sum_{e \in \beta(t_i)} f_i(e).$$

We restrict our attention to the case in which $f_1(e)$ and $f_2(e)$ are integral, and it is natural, in this case, to assume that $c(e)$ is integral too.

The two-commodity integral flow problem in directed networks (D2CIF) is defined as follows:

Input: A directed network N and two nonnegative integers R_1 and R_2, called the *requirements*.

Question: Are there integral flow functions f_1 and f_2 for N, for which $F_i \geq R_i$?

We shall show that D2CIF is NPC, even if all the edge capacities are 1; this is called the simple D2CIF.

Theorem 10.13: Simple D2CIF is NPC.

Proof: Let us show that SAT $\propto$ Simple D2CIF. The input I, to SAT, consists of clauses $C_1, C_2, \ldots, C_m$, each a subset of the set of literals $L = \{x_1, x_2, \ldots, c_n, \bar{x}_1, \bar{x}_2, \ldots, \bar{x}_n\}$. The structure of $f(I)$, the input to Simple D2CIF is as follows. For each variable x_i we construct a lobe, as shown in Fig. 10.7. Here p_i is the number of occurrences of x_i in the clauses, and q_i is the number of occurrences of $\bar{x}_i$. The lobes are connected in series: v_t^i is connected by an edge to v_s^{i+1}, s_1 is connected to v_s^1 and v_t^n to t_1. s_2 is connected by edges to all the vertices v_j^i and $\bar{v}_j^i$ where j is odd. In addition, there are vertices $C_1, C_2, \ldots, C_m$ and an edge from each to t_2. For the j-th occurrence of x_i ($\bar{x}_i$), there is an edge from v_{2j}^i ($\bar{v}_{2j}^i$) to vertex C_r, the clause in which it occurs. The requirements are $R_1 = 1$ and $R_2 = m$.

The first commodity must flow from s_1 to t_1 through the lobes; vertices $s_2, C_1, C_2, \ldots, C_m$ and t_2 cannot be used in this flow since there is no edge from the lobe to s_2, and there is no edge to return from $C_1, C_2, \ldots, C_m$ and t_2 to the lobes or to t_1. Thus, the one unit of the first commodity must use in each lobe either the upper or the lower path, but not both.

If the second commodity meets the requirement then $F_2 = R_2 = m$, and all the edges which enter t_2 are saturated. In this case, there is exactly one unit of flow, of the second commodity entering each C_k. If this unit of flow comes from the upper track of the i-th lobe, through the edge $v_{2j}^i \rightarrow C_k$, then clearly it uses also the edge $v_{2j-1}^i \rightarrow v_{2j}^i$ and the unit of the first commodity must use the lower track in this lobe.

Thus, if the answer to $f(I)$, with respect to D2CIF, is positive, then we can use the flows f_1 and f_2 to assign a satisfying assignment of the literals

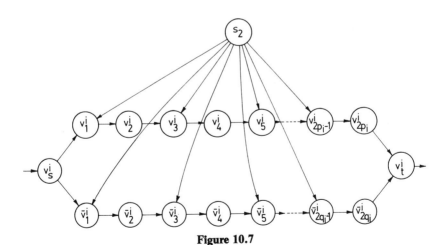

Figure 10.7

as follows: If the first commodity goes through the lower track of the i-th lobe, assign $x_i = T$, and if through the upper, $x_i = F$. In this case, the answer to I, with respect to SAT, is also positive.

Conversely, assume there is a satisfying assignment of the variables. If $x_i = T$, let the first commodity use the lower track in the i-th lobe; if $x_i = F$, use the upper track. Now, let ξ be a 'true' literal in C_k. If $\xi = x_i$ then the upper track is free of the first commodity and we can use it to flow one unit of the second commodity from s_2 to C_k; if $\xi = \bar{x}_i$, use the lower track. Finally, use the m edges entering t_2 to flow into it the m available units of the second commodity.

Q.E.D.

In the case of undirected networks, the graph $G(V, E)$ is undirected. The flow in the edges may be in either direction, and

$$f_i(u \xrightarrow{e} v) = -f_i(v \xrightarrow{e} u).$$

(Note that $c(u \xrightarrow{e} v) = c(e) \geq 0$.) Condition (1) on the edges, is changed to:

$$|f_1(u \xrightarrow{e} v)| + |f_2(u \xrightarrow{e} v)| \leq c(e).$$

Condition (2), that for each $v \in V\text{-}\{s_i, t_i\}$, the total flow of the i-th commodity entering v is equal to the total flow of the i-th commodity emanating from v, is now in the following form

$$\sum_{u \xrightarrow{e} v \in E} f_i(u \xrightarrow{e} v) = 0.$$

Note that in this equation v is fixed. Clearly,

$$F_i = \sum_{u \xrightarrow{e} t_i \in E} f_i(u \xrightarrow{e} t_i).$$

The undirected two-commodity integral flow problem (U2CIF) is defined similarly to D2CIF:

Input: An undirected network N and two nonnegative integers R_1 and R_2.

Question: Are there integral flow functions f_1 and f_2 for N, such that $F_i \geq R_i$?

Again we restrict our attention to simple networks, i.e., for all $e \in E$, $c(e) = 1$. We show that even for simple networks U2CIF is NPC.

Theorem 10.14: Simple U2CIF is NPC.

Proof: We show that Simple D2CIF $\propto$ Simple U2CIF.

First we change the digraph $G(V, E)$ of the directed network N, as follows: we add four new vertices $\bar{s}_1$, $\bar{s}_2$, $\bar{t}_1$ and $\bar{t}_2$ to serve as the two sources and sinks, respectively. We connect $\bar{s}_1$ to s_1 via R_1 parallel edges and t_1 to $\bar{t}_1$ via R_1 parallel edges. Similarly, $\bar{s}_2$ is connected to s_2 and t_2 to $\bar{t}_2$ via R_2 parallel edges in each case. Vertices s_1, s_2, t_1 and t_2 are now subject to the conservation rule and the requirements are the same. Clearly, the requirements can be met in the new digraph $G'(V', E')$ if and only if they can be met in the original one. Also, without loss of generality, we may assume that $R_1 + R_2 \leq |E|$, or obviously the requirements cannot be met.

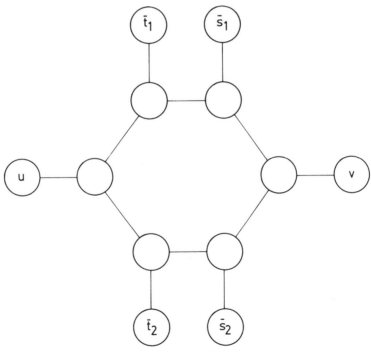

Figure 10.8

Thus, these changes can only expand the data describing the problem linearly.

We proceed to construct the undirected network from the new directed network N', as follows: Each edge $u \overset{e}{\to} v$ of G' is replaced by the construct shown in Fig. 10.8. (u or v may be any two vertices in V', including sources and sinks.) The vertices of the construct, which are unlabeled in Fig. 10.8, are new, and do not appear elsewhere in the graph.

The new requirements are $R_i' = R_i + |E'|$.

It remains to be shown that the requirements in N' can be met if and only if the new requirements can be met in the new undirected network $\tilde{N}$.

First assume that the requirements can be met in N'. Initially, flow one unit of each commodity through each of the constructs, as shown in Fig. 10.9. This yields $F_1 = F_2 = |E'|$. Next, if $u \overset{e}{\to} v$ is used in N' to flow one unit of the first commodity, then we change the flows in the corresponding construct as shown in Fig. 10.10. The case of the second commodity flowing through e in N' is handled similarly.

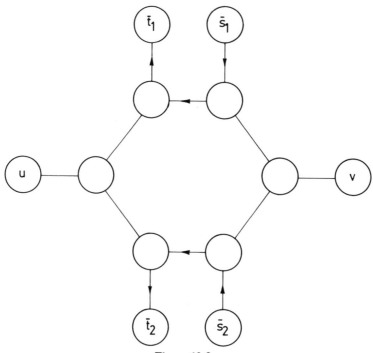

Figure 10.9

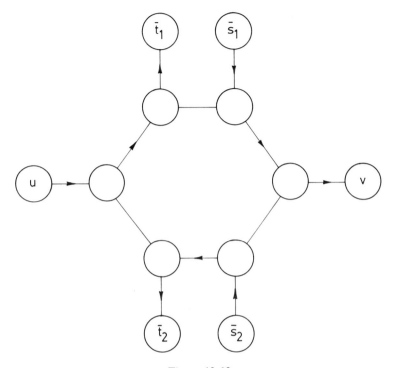

Figure 10.10

It is easy to see that R_1' and R_2' are now met in $\tilde{N}$.

Now assume we have a flow in the undirected graph satisfying the requirements R_1' and R_2'. Since the number of edges incident to $\bar{s}_i(\bar{t}_i)$ is R_i', all these edges are used to emanate (inject) i-th commodity flow from (into) $\bar{s}_i(\bar{t}_i)$. The flow through each edge-construct must therefore be in one of the following patterns:

(1) as in Fig. 10.9
(2) as in Fig. 10.10
(3) as in Fig. 10.10, for the second commodity.

We can now use the following flows through $u \xrightarrow{e} v$, in N': If in e's construct, in $\tilde{N}$, we use pattern (1) then $f_1(e) = f_2(e) = 0$. If it is pattern (2), then $f_1(e) = 1$ and $f_2(e) = 0$, etc. Clearly, this defines legal flows in N' which meet its requirements.

Q.E.D.

It is easy to see that the multicommodity integral flow problems, as we have defined them in the reductions, are easily reducible to the version in which we have only one total requirement, i.e. $F_1 + F_2 \geq R$. Thus, the latter versions are NPC too. Also, the completeness of $m > 2$ commodity integral flow problem follows as an immediate consequence.

PROBLEMS

10.1 Show that IND $\propto$ CLIQUE without using Cook's theorem (Theorem 9.1) or any of its consequences.

10.2 Show that the problem of finding a minimum independent set which is also a vertex cover is solvable in polynomial time.

10.3 A set $S \subseteq V$, in $G(V, E)$, is called a *dominating set* if for every $v \in V - S$, there exists an edge $u - v$ in G, such that $u \in S$. Formulate the minimum dominating set problem as a decision problem and prove that it is NPC. (Hint: One way is to show a reduction from VC. For every edge $u - v$ add a new path $u - x - v$ in parallel).

10.4 Formulate the problem of finding a minimum independent set which is also dominating, as a decision problem and prove its NP-completeness. (Hint: Use 10.3. Duplicate vertices. The set of duplicates is independent. Add to each duplicate a path of length 2.)

10.5 Formulate the traveling salesman problem for undirected graphs, where a circuit is required but it may pass through vertices more than once, as a decision problem, and prove that it is NPC.

10.6 Show that there is a polynomial algorithm for finding a circuit C, as in 10.5, whose length $l(C)$ satisfies $l(C) \leq 2 \cdot l(T)$, where $l(T)$ is the length of a minimum spanning tree T of the graph. Prove also that every such circuit C satisfies $l(C) > l(T)$, if all edge lengths are positive.

10.7 Appel and Haken [6] proved that every plane graph is 4-colorable. (This is the famous four color theorem.) Thus, 4-colorability of plane graphs is polynomial (why?). Prove that for every $k \geq 3$, k-colorability of graphs in general is NPC.

10.8 The following is called the partition of a graph into cliques problem:

Input: A graph $G(V, E)$ and a positive integer k.

Question: Can V be partitioned into k sets V_1, V_2, ..., V_k such that the subgraph of G, induced* by each V_i, is a clique?

Prove that this problem is NPC.

10.9 Consider the following problem on digraphs:

Input: A strongly connected digraph $G(V, E)$ and a positive integer k.

Question: Is there a subset of edges E' such that $|E'| \leq k$ and $G'(V, E')$ is strongly connected?

Prove that this problem is NPC.

10.10 As we have seen in Chapter 1, the shortest path problem is polynomially solvable. Show that even the following "simple" longest path problem is NPC.

Input: A graph $G(V, E)$, two vertices s and t and a positive integer k.

Question: Is there a simple path in G with endpoints s and t and length (measured by the number of edges) greater than or equal to k?

10.11 Prove that the Hamilton path (HP) problem for undirected graphs, remains NPC even if we restrict the graphs to be bipartite. (Hint: A minor change in the proof of Theorem 10.5.)

10.12 Prove that the Hamilton path problem for undirected graphs remains NPC (using Karp's definition) even if s and t are not specified.

10.13 The following is called the degree restricted spanning tree problem:

Input: A graph $G(V, E)$ and a positive integer k.

Question: Is there a spanning tree for G in which for every vertex v, $d(v) \leq k$?

Prove that the problem is NPC even if $k \geq 2$ is fixed and not part of the input. (Hint: First use 10.12 to prove for $k = 2$.)

10.14 Consider the following network communication problem. $G(V, E)$ is a graph, a vertex $v_0 \in V$ is called the center. Each vertex $v \in V\text{-}\{v_0\}$ is assigned a requirement $r(v) \geq 0$ (the quantity to be shiped to v_0).

---

*The subgraph of $G(V, E)$ induced by $S \subseteq V$, is the graph (S, E'), where E' is the subset of E which includes all the edges whose endpoints are in S.

Each edge e is assigned a capacity $c(e) > 0$ and a cost $k(e) \geq 0$. The problem is to find a minimum cost spanning tree T which satisfies the following condition: If $U(e)$ is the set of vertices such that the paths from v_0 to them go through e then $\sum_{v \in U(e)} r(v) \leq c(e)$. The cost of T is $\sum_{e \in T} k(e)$. Form the corresponding decision problem and prove that it is NPC. (Hint: A construction similar to the one in the proof of Theorem 10.10 may be used.)

10.15 The following may be called the directed Steiner tree problem (with no edge lengths):

Input: A digraph $G(V, E)$, a set of vertices X, a vertex $r \in X$, a positive integer k.

Question: Is there a directed tree $T(W, F)$ in G, with root r, such that $X \subseteq W \subseteq V$ and $|F| \leq k$? Prove that the problem is NPC.

10.16 Prove that the Steiner tree problem remains NPC even if the following restrictions are imposed: G is bipartite and the set X, to be connected, is one of its parts, i.e.: $G = (X, Y, E)$.

10.17 The following is called the minimum edge-deletion bipartite subgraph problem (MINEDB):

Input: A graph $G(V, E)$ and a positive integer k.

Question: Is there a subset of edges, E', such that the subgraph $(V, E - E')$ is bipartite and $|E'| \leq k$? Prove that MINEDB is NPC. (Hint: MAXC $\propto$ MINEDB.)

10.18 The following is called the minimum cut into equal-sized subsets problem (MINCES):

Input: A graph $G(V, E)$, two vertices s and t and a positive integer k.

Question: Is there a partition of V into S and $\bar{S}$ such that

$$|S| = |\bar{S}|, \ s \in S, \ t \in \bar{S} \text{ and } |\{u - v | u \in S, v \in \bar{S}\}| \leq k?$$

Prove that MINCES is NPC. (Hint: Garey et al [3], use MAXC $\propto$ MINCES. If I consists of $G'(V', E')$ and k', $f(I)$ consists of $G(V, E)$, s, t and k as follows:

$$V = V' \cup \{x_1, x_2, \ldots, x_n\} \text{ where } n = |V'|$$
$$E = \{u - v | u \neq v \text{ in } G'\}$$

$$s = x_1$$

$$t = x_n$$

$$k = n^2 - k'.)$$

10.19 Prove that the problem of whether a maximum flow in a network can be constructed by using only k augmenting paths (using the Ford and Fulkerson algorithm) is NPC. (Hint: Consider the network shown below and relate it to the Knapsack problem. See Problem 9.4.)

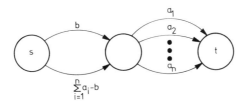

10.20 The following network flow problem is called the integral flow with bundles problem (IFWB). As in the maximum flow problems, there is a digraph $G(V, E)$, a vertex s, assigned as the source and a vertex t, assigned as the sink. The bundles are subsets of edges, B_1, $B_2, \ldots, B_k$. There are bundle capacities $c_1, c_2, \ldots, c_k$, and the flow f must satisfy the condition that for each B_i, $\sum_{e \in B_i} f(e) \leq c_i$. In addition, for every vertex $v \in V - \{s, t\}$, the incoming flow must equal the outgoing flow. The question is whether there is a flow which meets the requirement R. Prove that IFWB is NPC. (Hint: May be proved by IND $\propto$ IFWB [12].)

REFERENCES

[1] Garey, M. R., and Johnson, D. S., *Computers and Intractability, A Guide to the Theory of NP-Completeness*, Freeman, 1979.

[2] Stockmeyer, L. J., "Planar 3-Colorability is NP-Complete", SIGACT News, Vol. 5, #3, 1973, pp. 19–25.

[3] Garey, M. R., Johnson, D. S., and Stockmeyer, L. J., "Some Simplified NP-Complete Graph Problems", Theor. Comput. Sci., Vol. 1, 1976, pp. 237–267.

[4] Kautz, W. H., Levitt, K. N., and Waksman, A., "Cellular Interconnection Arrays", IEEE Trans. Computers, Vol. C-17, 1968, pp. 443–451.

[5] Even, S., *Algorithmic Combinatorics*, Macmillan, 1973. (See Section 1.4.)

[6] Appel, K., and Haken, W., "Every Planar Map is Four Colorable", Bull. Amer. Math. Soc., Vol. 82, 1976, pp. 711-712.

[7] Karp, R. M., "Reducibility among Combinatorial Problems", in R. E. Miller and J. W. Thatcher (eds.), *Complexity of Computer Computations*, Plenum Press, 1972, pp. 85-104.

[8] Garey, M. R., Graham, R. L., and Johnson, D. S., "The Complexity of Computing Steiner Minimal Trees", *SIAM J. Appl. Math.*, Vol. 32, 1977, pp. 835-859.

[9] Even, S., and Shiloach, Y., "NP-Completeness of Several Arrangement Problems", Technical Report #43, Dept. of Comp. Sci., Technion, Haifa, Israel, Jan. 1975.

[10] Even, S., Itai, A., and Shamir, A., "On the Complexity of Timetable and Multicommodity Flow Problems", *SIAM J. Comput.*, Vol. 5, #4, Dec. 1976, pp. 691-703.

[11] Itai, A., "Two Commodity Flow", *J. ACM*, Vol. 25, #4, Oct. 1978, pp. 596-611.

[12] Sahni, S., "Computationally Related Problems", *SIAM J.* on Comput., Vol. 3, 1974, pp. 262-279.

INDEX

DATE DUE

DATE DUE		
MAY 2 7 1996		
MAR 2 6 1997		
APR 1 6 1997		
MAY 1 0 1997		
MAR 16 1998		
APR 1 7 1998		
MAY 1 8 1998		
GAYLORD No. 2333		PRINTED IN U.S.A.